AF327149

Protein Structure–Function Relationships in Foods

This monograph is dedicated to Professor Shuryo Nakai on the occasion of his retirement from the Department of Food Science, University of British Columbia, where he has had a long and distinguished career. He was one of the first to introduce and advance the concept of hydrophobicity, to identify its importance in food protein structure–function relationships, and to promote the use of multivariate techniques in food protein structure–function studies. In the words of Isaac Asimov: 'It is hard to describe the exact route to scientific achievement, but a good scientist doesn't get lost as he travels it.' Here is to one of the great Food Science Travellers of our time.

Protein Structure–Function Relationships in Foods

Edited by

R. Y. YADA
Department of Food Science
University of Guelph
Guelph
Ontario

R.L. JACKMAN
ORTECH Corporation
Mississauga
Ontario

and

J.L. SMITH
Ault Foods Ltd.
London
Ontario

BLACKIE ACADEMIC & PROFESSIONAL

An Imprint of Chapman & Hall

London · Glasgow · Weinheim · New York · Tokyo · Melbourne · Madras

Published by
Blackie Academic and Professional, an imprint of Chapman & Hall,
Wester Cleddens Road, Bishopbriggs, Glasgow G64 2NZ

Chapman & Hall, 2–6 Boundary Row, London SE1 8HN, UK

Blackie Academic & Professional, Wester Cleddens Road, Bishopbriggs, Glasgow G64 2NZ, UK

Chapman & Hall GmbH, Pappelallee 3, 69469 Weinheim, Germany

Chapman & Hall Inc., One Penn Plaza, 41st Floor, New York NY 10119, USA

Chapman & Hall Japan, Thomson Publishing Japan, Hirakawacho Nemoto Building, 6F, 1–7–11 Hirakawa-cho, Chiyoda-ku, Tokyo 102, Japan

DA Book (Aust) Pty Ltd, 648 Whitehorse Road, Mitcham 3132, Victoria, Australia

Chapman & Hall India, R. Seshadri, 32 Second Main Road, CIT East, Madras 600 035, India

First edition 1994

© 1994   Chapman & Hall

Typeset in 10/12pt Times by Florencetype Ltd, Kewstoke, Avon
Printed in Great Britain at the Cambridge University Press, Cambridge

ISBN   0 7514 0186 2

A catalogue record for this book is available from the British Library
Library of Congress Catalog Card Number: 94 – 70121

♾ Printed on acid-free text paper, manufactured in accordance with ANSI/NISO Z39.48–1992 (Permanence of Paper)

Preface

Food proteins constitute a diverse and complex collection of biological macro-molecules. Although contributing to the nutritional quality of the foods we consume, proteins also act as integral components by virtue of their diverse functional properties. The expression of these functional properties during the preparation, processing and storage of foods is largely dictated by changes to the structure or structure-related properties of the proteins involved. Therefore, germane to the optimal use of existing and future food protein sources is a thorough understanding of the nature of the relationships between structure and function. It is the goal of this book to aid in better defining these relationships. Two distinct sections are apparent: firstly, those chapters which address structure–function relationships using a variety of food systems as examples to demonstrate the intricacies of this relationship, and secondly, those chapters which discuss techniques used to either examine structural parameters or aid in establishing quantitative relationships between protein structure and function.

The editors would like to thank all contributors for their assistance, co-operation and, above all, their patience in putting this volume together, and the following companies/organizations for their financial support without which it would not have been the success it was: Ault Foods Limited, Best Foods Canada Limited, Natural Sciences and Engineering Research Council of Canada, Ontario Ministry of Agriculture and Food, Quest International Canada Inc., and University of Guelph.

R.Y.Y.
R.L.J.
J.L.S.

Series foreword

The 8th World Congress of Food Science and Technology, held in Toronto, Canada, in 1991 attracted 1400 delegates representing 76 countries and all five continents. By a special arrangement made by the organisers, many participants from developing countries were able to attend. The congress was therefore a most important international assembly and probably the most representative food science and technology event in that respect ever held. There were over 400 poster presentations in the scientific programme and a high degree of excellence was achieved. As in previous congresses, much of the work reported covered recent research and this will since have been published elsewhere in the scientific literature.

In addition to presentations by individual researchers, a further major part of the scientific programme consisted of invited papers, presented as plenary lectures by some of the leading figures in international food science and technology. They addressed many of the key food issues of the day including advances in food science knowledge and its application in food processing technology. Important aspects of consumer interest and of the environment in terms of a sustainable food industry were also thoroughly covered. The role of food science and technology in helping to bring about progress in the food industries of developing countries was highlighted.

This book is part of a series arising from the congress and including full bibliographical details. The series editors are Professor M. A. Tung of the Technical University of Nova Scotia, Canada; and Dr G. E. Timbers of Agriculture Canada, Ottawa, Canada. The book presents some of the most significant ideas which will carry food science and technology through the nineties and into the new millennium. It is therefore essential reading for anyone interested in the subject, including specialists, students and general readers. IUFoST is extremely grateful to the organisers from the Canadian Institute of Food Science and Technology for putting together a first class scientific programme and we welcome the publication of this book as a permanent record of the keynote papers presented at the congress.

Dr D. E. Hood
(President, International Union of Food Science & Technology)

Contributors

T. Aishima Kikkoman Corporation, Noda, Chiba-ken, Japan.

G.L. Catignani Southeast Dairy Foods Research Center, Department of Food Science, North Carolina State University, Raleigh, North Carolina, USA.

S.X. Chen Southeast Dairy Foods Research Center, Department of Food Science, North Carolina State University, Raleigh, North Carolina, USA.

F. Finardi Filho Department of Food Science and Nutrition, São Paulo University, São Paulo, Brazil.

K.L. Fuller Department of Food Science, Acadia University, Wolfville, Nova Scotia, Canada.

A.M. Hermansson SIK—The Swedish Institute for Food Research, Göteborg, Sweden.

M. Hirotsuka Fuji Oil Co. Ltd., Osaka-fu, Japan.

M.F. Ho Department of Food Science and Technology, University of California, Davis, California, USA.

R.L. Jackman ORTECH Corporation, Mississauga, Ontario, Canada.

J.E. Kinsella *formerly of* College of Agriculture and Environmental Science, University of California, Davis, California, USA.

F. Lajolo Department of Food Science and Nutrition, São Paulo University, São Paulo, Brazil.

E. Li-Chan Department of Food Science, University of British Columbia, Vancouver, British Columbia, Canada.

S. Nakai Department of Food Science, University of British Columbia, Vancouver, British Columbia, Canada.

S. Oh Southeast Dairy Foods Research Center, Department of Food Science, North Carolina State University, Raleigh, North Carolina, USA.

D. Osuga Department of Food Science and Technology, University of California, Davis, California, USA.

L.G. Phillips Department of Food Science, Cornell University, Ithaca, New York, USA.

D.J. Rector A.E. Staley Manufacturing Co., Galesburg, Illinois, USA.

S.G. Roscoe Department of Food Science, Acadia University, Wolfville, Nova Scotia, Canada.

J.L. Smith Ault Foods Ltd., London, Ontario, Canada.

H.E. Swaisgood Southeast Dairy Foods Research Center, Department of Food Science, North Carolina State University, Raleigh, North Carolina, USA.

A. van der Schaaf Department of Food Science and Technology, Agricultural University, Wageningen, The Netherlands.

J.R. Whitaker Department of Food Science and Technology, University of California, Davis, California, USA.

R.Y. Yada Department of Food Science, University of Guelph, Guelph, Ontario, Canada.

X. Yin CIBA-Geigy Ltd, Basle, Switzerland.

Contents

4 Control of polyphenoloxidase activity using a catalytic mechanism   62
D. OSUGA, A. VAN DER SCHAAF and
J.R. WHITAKER

5 Naturally occurring α-amylase inhibitors:
Structure/function relationships   89
M.F. HO, X. YIN, F.F. FILHO, F. LAJOLO and
J.R. WHITAKER

6 Application of multivariate analysis in studies of food
protein functions   120
S. NAKAI, T. AISHIMA and R.Y. YADA

1 Physicochemical properties of proteins: Texturization via gelation, glass and film formation

J.E. KINSELLA*, D.J. RECTOR AND L.G. PHILLIPS

Abstract

Proteins because of their dynamic structures and amphiphilic nature possess varying functional properties. This chapter discusses the relationship between molecular structure and functional properties, i.e. gelation, glass transition and foaming properties of proteins. The multitude of possible interactions underscore the challenge of completely describing and controlling all the factors and interactions that control gelation and foam formation. The dynamic nature and flexibility of proteins indicate the possibility of further enhancing the foaming properties of proteins.

This research indicates the need for continuing research that attempts to interrelate the chemistry, structure, conformation and physicochemical properties of food proteins. More emphasis needs to be placed on understanding how varying processing conditions affect the flexibility and the interaction of proteins for the purpose of controlling functionality. Developing new information will expand uses for food macromolecules and extend product development.

1.1 Introduction

The food industry has changed from a commodity handling industry toward a market-driven consumer products industry. The contemporary consumer is increasingly demanding food products that are compatible with a busy, healthy lifestyle that include convenience, balanced calories and nutrients, safe and more wholesome (less saturated fatty acids and cholesterol) with consistent high quality, appropriate portioning and attractive packaging. The quality attributes, i.e. flavor, odor, color, taste, texture and mouthfeel are expected (Kinsella and Phillips, 1989). More-fabricated food products will be increasingly manufactured, placing a premium on ingredients with versatile but consistent functional properties and that are compatible with automated formulation or fabrication.

*All correspondence should be directed to L. G. Phillips.

These developments have demonstrated the need for a range of functional ingredients, in particular low-calorie and structure-forming macromolecules such as polysaccharides and proteins.

Proteins represent a most important class of functional ingredients because they possess a range of dynamic functional properties (Table 1.1); they show versatility during processing, they can form networks and structures and they

Table 1.1 Functional properties of proteins in foods

General property	Functional criteria
Organoleptic	Color, flavor, odor
Kinaesthetic	Mouthfeel, texture, smoothness, grittiness
Hydration	Solubility, wettability, water sorption, swelling, thickening, gelling, synersis, viscosity, gelation
Surface	Emulsification, foaming, film formation
Rheological/textural	Elasticity, cohesiveness, chewiness, adhesiveness, network formation, aggregation, dough formation, texturizability, extrudability

provide essential amino acids, i.e. they fulfil functional and nutritional requirements. In addition, they interact with other components and improve quality attributes of foods.

Many sources of functional proteins are used in foods. The bulk of these are used in products that are relatively tolerant of variability in the ingredient proteins; however, with refinements in product formulae and with automated formulation the food industry is becoming more demanding not only for compositional but also for functional specifications of ingredient proteins. This has emphasized the need for standard methods to describe quantitatively the functional properties of proteins for applications and also help elucidate structure–function relationships (Kinsella, 1976; Phillips *et al.*, 1990a).

The commodity (e.g. dairy, egg, soy) industry will depend increasingly on the food industry as a market for functional ingredients; hence, it must ensure that the varied and exacting specifications of the food industry are met and that methods of preparation (e.g. separation, dehydration, fractionation) of functional ingredients are adequate. Extensive research is warranted to show the unique functional attributes of individual proteins, protein blends or modified protein. In this context, much more information concerning the relationships between the physical properties of proteins and their functional behavior under different conditions is needed. Furthermore, the particular physical properties that meet the functional requirements in particular food applications must be elucidated so that rational decisions can be made in selecting the best proteins for specific applications or determining what modifications are required for improving a particular function.

Different applications require quite different functional properties and many products require an array of properties. In some cases, products depend on changes in properties during actual processing or preparation. For example, during foam or gel formation and stabilization, some molecular unfolding and subsequent protein–protein interactions must occur. In reformed meats, adhesion and binding are important and adhesive proteins that can function without added salt are needed. In comminuted meat products, solubility, viscosity, emulsifying capacity, water-holding capacity and gelation are required to ensure good texture, shape retention, cutting characteristics and smooth mouthfeel. Few single proteins possess the appropriate range of properties required to perform all of these functions, therefore a mix of proteins is usually required. In foams or emulsions the protein(s) should possess good interfacial activity and form strong cohesive, elastic films (Kinsella and Phillips, 1989). To achieve this, a protein must perform a sequence of functions that few individual proteins can achieve satisfactorily.

Knowledge of the physicochemical characteristics required for particular uses is therefore important. In general, polypeptides that facilitate interactions and create a balance of attractive and repulsive forces are desirable. In foams and emulsions, amphiphilic and flexible molecules that orient readily at an oil–water interface with maximal protein–protein interaction are required to form a strong film. Some of the properties of proteins that relate to gelation, film-forming and foaming properties are emphasized in this chapter.

1.2 Gelation

Gelation and structure-formation are important functional properties of food proteins in many fabricated and natural food products, e.g. gelatin, egg white and comminuted meat products (Kinsella, 1982, 1984a,b). In each of these products, proteins contribute in varying degrees to the solid or elastic properties of the food by formation of an orderly, three-dimensional network of associated or aggregated protein molecules that are capable of physically entrapping large amounts of water within the matrix (Hermansson, 1979).

The formation of a gel from protein is apparently a two-step process. The first step involves a change in conformation (usually heat-induced) or partial denaturation of the protein molecules. As denaturation proceeds, the viscosity of the dispersion increases owing to an increase in molecular dimensions of the unfolding proteins (Catsimpoulas and Meyer, 1970). This is followed by a gradual association or aggregation of the individual denatured proteins (Ferry, 1948). During the association step, there is an exponential increase in viscosity as the material approaches a continuous network. This dispersion of protein aggregates then begins to display some of the characteristics of an elastic solid, i.e. the storage modulus (G') of the solution increases. This second step should be slow, relative to the first, so that a well-organized gel network is formed. If the second step occurs too quickly, a random network (i.e. a coagulum) that is unable to hold water is formed and syneresis occurs. A critical balance between attractive and repulsive forces must also be present for successful network formation and stabilization (Hermansson, 1979). If attractive forces predominate, a coagulum is formed and water is expelled from the gel matrix. If repulsive forces predominate, no network will be formed (Kinsella, 1984a).

The type and properties of gels are sensitive to many factors, including protein concentration, pH, type of salt and salt concentration (Mulvihill and Kinsella, 1988). Gelation may occur during heating or upon cooling depending on the protein and conditions of gelation. A thermoset gel is formed upon heating, and thereafter cannot be remelted without destroying the primary structure of the original protein molecules (Rodriguez, 1982; Young, 1983). The process involves the formation of an elastic solid, a permanently cross-linked three-dimensional solid network as exemplified by vulcanized rubber, soy, egg white and traditional heat-induced whey protein gels (Clark *et al.*, 1982).

A thermoplastic gel, as the name implies, melts and flows upon heating (Rodriguez, 1982; Young, 1983). The noncovalent bonds in a thermoplastic gel progressively melt (break) upon heating. The most familiar example of a thermoplastic food gel is gelatin. Other examples include peanut arachin (Kella *et al.*, 1980) and lysozyme (Clark and Lee-Tuffnel, 1986).

Whey proteins represent a major source of good-quality functional proteins with many potential uses in foods (Kinsella, 1984a,b). There have been many studies on the properties of and factors affecting the heat-induced thermoset gelation of various whey proteins (Mulvihill and Kinsella, 1987). Whey proteins are also capable of forming thermoplastic (reversible) gels and capable of controlling the viscosities of solutions when heated under the appropriate conditions. Several factors affect the types, characteristics and physical properties of whey protein gels and solutions obtained by heating dispersions of whey protein. The type of material obtained by heat-treating whey protein dispersions is sensitive to several factors including pH of the solution, calcium concentration and overall protein concentration (Rector *et al.*, 1989).

Protein concentration is a very important factor in determining the type of gel and final gel characteristics. Dispersions of dialyzed whey protein isolates (DWPI) heated at 90°C in the concentration range of 9–10.5% protein (pH 6.5–8.0) formed reversible gels when cooled to 6°C. Pure β-lactoglobulin also formed reversible gels under these conditions at concentrations of 8.0–9.0% protein. Both these gels re-melted into viscous liquids upon reheating (Rector *et al.*, 1989). Gels produced at higher protein concentrations had higher melting temperatures, were firmer and quicker setting. A heated whey protein dispersion (8.5% protein) took 168 h to set into a very weak gel at 6°C. A 90% dispersion set within 36 h and at protein concentrations above 9.5% the gels readily set within 12 h at 6°C.

The reversibility of gelation was reflected in the changes in viscoelastic parameters of DWPI gels. The loss modulus, G" (a viscous energy dissipation term) increased with time, although the rate of the increase declined during the later stages of the experiments. The storage modulus, G' (a term representing energy stored elastically in the protein network) increased at a constant and more rapid rate for up to 7 h (Figure 1.1). In all experiments, the storage modulus returned

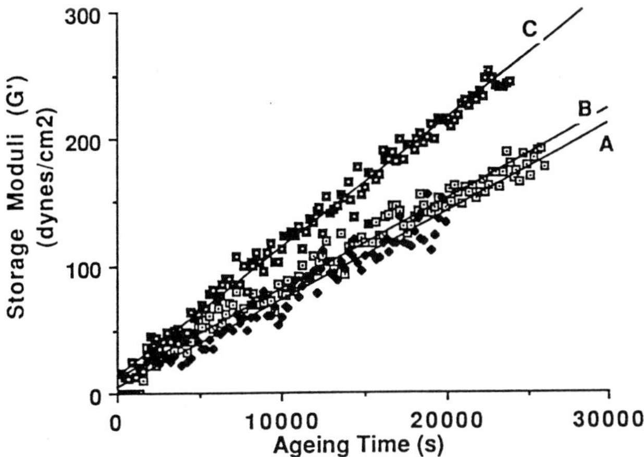

Figure 1.1 The progressive increase in the storage modulus (G') of melted dialyzed whey protein isolate (DWPI) gels at pH 8.0 while being held on the plate of the mechanical spectrometer at 8°C for one (A), two (B) and three (C) cycles of melting and gelation. Values of G' returned to zero for all cycles after heating (from Rector *et al.*, 1989).

A. G' = 4.62 + 6.84e-3t r^2 = 0.953.
B. G' = 12.26 + 6.96e-3t r^2 = 0.970.
C. G' = 13.05 + 1.01e-2t r^2 = 0.989.

to 0 upon heating at 90°C, indicating completely reversible gelation. The rate of gel development or setting increased after each successive reheating, indicating that there was a more rapid development of a gel network. The increased rates of setting after each successive melting cycle may have been the result of a continued thiol-disulfide interchange reaction and, concomitantly, more extensive unfolding of the protein molecules. The reversible whey gels possessed very weak mechanical properties compared with gelatin. The largest value of G' obtained (after three reheating cycles) was 240 dynes/cm² after 7 h of aging. By extrapolation, the value of G' increased to 877 dynes/cm² after 24 h of aging. The 10% DWPI gels after 24 h of aging would not be self-supporting. Typical values of G' for 5% gelatin gels (MW = 60kD) are about 33 000 and 42 000 dynes/cm² after 7 h and 24 h of aging, respectively (te Nijenhuis, 1981a,b). The loss tangent (G"/G') generally declined with aging as more energy from the deformation of the material was stored elastically in the gel network.

Melting temperatures also increased with increasing protein concentration and aging time (Table 1.2). After 66 h at 6°C, the melting temperature seemed to have nearly reached a constant value. Similar results have been observed for gelatin gels by measuring the increase in G' with time. The increase of G' of the gelatin gels has been explained as a slow approach to an equilibrium configuration or concentration of hydrogen-bonded cross-links in the reversible gel network (te Nijenhuis, 1981a,b). The reversible whey protein gels seem to

Table 1.2 Effects of pH, protein concentration and age on the melting temperatures of gels made from dialyzed whey protein isolate[a]

pH of solution	Concentration (%)	Temperature of melting (°C)		
		17 h	41 h	66 h
7.5	10.25	37.7	43.9	47.3
7.5	10.35	39.3	44.1	47.2
7.5	10.50	41.1	46.4	49.5
7.5	10.60	51.1	53.8	55.0
7.5	10.75	55.0	57.2	58.4
8.0	9.50	24.5	31.1	34.3
8.0	9.75	31.1	36.9	39.8
8.0	10.00	37.6	43.3	46.9
8.0	10.25	44.4	49.5	51.5
8.0	10.50	52.1	56.2	57.8

[a]Source: Rector *et al.* (1989).

be undergoing a similar process. By assuming that an equilibrium had been reached between the actual number of formed cross-links and the total number of possible cross-links the change in enthalpy and entropy for the formation of the gel network was calculated on the basis of the van 't Hoff equation (Eldridge and Ferry, 1954; Atkins, 1982). The maximum enthalpy of formation of the cross-links for reversible whey protein isolate gels was calculated to be −858 kcal/mole after aging 66 h at 6°C. The enthalpies at shorter aging times were lower because the system was still approaching an equilibrium cross-link density. These values are much lower than the 49–73 kcal/mole of cross-links obtained for gelatin by the same method (Eldridge and Ferry, 1954). A value of 1.2 kcal/mole of cross-links was obtained for reversible peanut arachin gels (Kella, 1987). The high enthalpy of melting for gelatin has been attributed to the large numbers of hydrogen bonds involved in the formation of each noncovalent cross-link. Gelatin also possesses a unique molecular structure that tends to form the helical structures involved in cross-linking. The low enthalpies involved in formation of DWPI reversible gels are probably the result of the balance between the attractive and repulsive forces stabilizing the reversible network, and the molecular structure of the whey proteins that tends to form globular tertiary structures. The general mechanism involved in reversible whey protein gelation is the formation of large polymers of proteins. These polymers are unable to form a continuous network at concentrations below 10.5%. The large polymers participate in noncovalent bonding (most likely hydrogen bonding) at low temperatures. The hydrogen bonds are broken at higher temperatures.

Whey protein isolates also form thermoset gels when heated at 90°C for 15 min at concentrations above 10.5% (pH 8.0). The strength or hardness of these gels increases with the square of protein concentration as has been demonstrated for pure β-lactoglobulin (Rector, 1992). The general mechanism for formation of thermoset whey protein gels involves the formation of a covalent network of intermolecular disulfide bonds. The formation of these intermolecular disulfide bonds also occurred spontaneously at room temperature and at a much lower temperature when urea was included in the whey protein dispersions under the same pH conditions (Xiong and Kinsella, 1990). The initial network of disulfide bonds is reinforced by the formation of noncovalent interactions to provide additional cross-links (Beveridge *et al.*, 1984). Thus, heating and/or urea causes dissociation of the β-lactoglobulin dimers (Kella and Kinsella, 1988) with some conformational changes that expose the free thiol groups, which then engage in thiol–disulfide interchange to form the three-dimensional matrix. The rate of exposure, which is pH-dependent, can affect the firmness of the gel network.

The pH of the protein solution also affects greatly the gelling characteristics of whey proteins. Reversible gels were formed only in the pH range between 6.5 and 8.5. Dispersions of 10% DWPI formed opaque coagula of gels at pH 5.0–6.0. The same solutions at pH 1.5–3.5 formed thermoset gels. These gels were very weak and pasty in consistency. This consistency was very different from the more elastic-like gels formed in the pH range of 6.5–8.5. At concentrations of less than 8.0% DWPI, viscous solutions were formed. As previously stated, the thermoset and thermoplastic forms of whey protein gelation occur as a result of the thiol-disulfide interchange reaction at pH above 6.5. The thiol-disulfide interchange reaction was not involved in the gelation of the acid whey protein solutions. At pH 3.0, the sample contained very few polymerization products after being heated for 15 min at 90°C. The pH 6.0 sample contained a high degree of polymerization after 5 min and almost complete conversion to high molecular weight polymers after 15 min heating at 90°C. These results are consistent with the polymerization of the whey protein solutions observed at pH 8.0. The gels formed at an acid pH are also weak and nonelastic. These characteristics are different from the highly elastic, basic pH gels. The fact that the acidic gels did not melt at higher temperatures and had formed no covalent cross-links may indicate that the gel network was stabilized by hydrophobic interactions. Haque and Kinsella (1988) demonstrated that heating β-lactoglobulin and κ-casein in solution resulted in an increase in hydrophobic-type interaction. Heating of β-lactoglobulin solutions at pH 3.0 has also been shown to result in an increase in the surface hydrophobicity as determined by ANS binding (Phillips, 1992).

In addition to the above factors, the amount of time that the whey proteins have been stored as a dry powder also greatly affects their gelling properties (Figure 1.2). The data shown in Figure 1.2 illustrate the deterioration of gelling properties of whey proteins with time when stored at 80°C. The hardness of both whey protein and β-lactoglobulin gels decreased with increasing storage time at all gelation temperatures tested (75–95°C). The decrease in overall gelation quality appeared to be the result of a nondisulfide type of covalent cross-linking that was responsible for the polymerization of the individual proteins as determined by SDS-PAGE. This reaction apparently followed second-order kinetics (Rector, 1992). The rate constants were determined for temperatures of 40–80°C. It should be noted that there was a continued polymerization at 40°C, although at a much

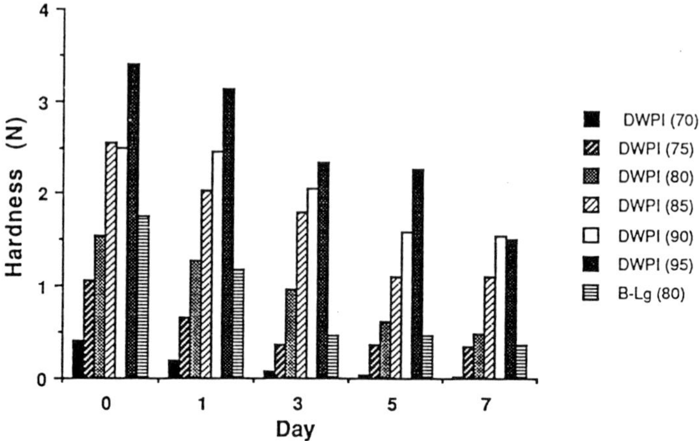

Figure 1.2 The progressive decrease in hardness of gels made from dialyzed whey protein isolate (DWPI) and β-lactoglobulin powder held at 80°C for up to 7 days prior to gelation. Gels were formed by heating 12% dialyzed whey protein isolate solutions and 10% β-lactoglobulin solutions (pH 8.0) for 30 min at the specified temperatures (70–95°C), cooling the gel at 4°C for 12 h and determining hardness (from Rector *et al.*, 1989).

slower rate. The polymerization has also been observed in protein samples that were stored for 2 years at 4°C. An Arrhenius plot was used to quantify rates for temperatures other than 80°C. From these data, it was estimated that 19% of the original monomeric proteins would be converted to higher molecular weight materials after 1 year of dry storage at 25°C.

Whey protein isolate dispersions (pH 8.0) at concentrations of less than 9.5% protein form viscous liquids when heated at 90°C for 15 min. The viscosities of these heated solutions depend on protein concentration, heating temperature and time. The viscosities of heated whey solutions at pH 8.0 dramatically increase with increasing protein concentration (Rector *et al.*, 1991). A 9% DWPI dispersion (pH 8.0) heated for 15 min at 90°C has a viscosity of 10.3 cp. The viscosity of an unheated 9% sample of the same material is 1.8 cp. A 7% and 5% solution treated under the same conditions had viscosities of 6.0 and 1.9 cp, respectively (Figure 1.3). The increased viscosities observed under these conditions were the result of protein polymerization. In addition to creating large molecules, polymerization also acted to eliminate intramolecular disulfide bonds. The extent of the individual protein polymerization was highly dependent on concentration. Elimination of these intramolecular bonds with dithiothreitol (DTT) increased the dispersion viscosity. The highest viscosity for a heated 7% whey protein isolate dispersion was observed following treatment with DTT. A 9% whey protein dispersion heated at 90°C with DTT formed a thermoplastic gel with a melting point of approximately 85°C. These data suggest that the intramolecular disulfide bonds in native β-lactoglobulin constrain the molecular conformations

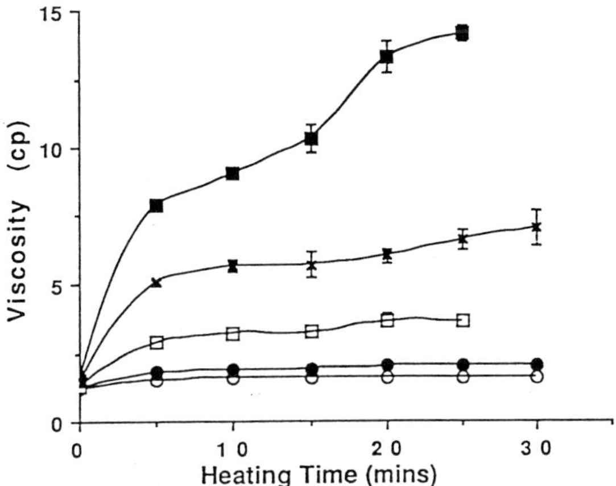

Figure 1.3 Effects of heating for various times at 90°C on the viscosity of dialyzed whey protein isolate (DWPI) dispersions (5–9%, pH 8.0). ■, 9%; × 8%; □, 7%; ●, 6%, ○, 5%.

of the whey proteins and, therefore, greatly decrease their ability to form non-covalently cross-linked gel networks.

A possible application of the reversible whey protein gels or the viscous whey protein solutions is the use of the material in protein glasses or plasticized glasses. By pretreating the proteins to form reversible gels, a wide range of material strengths can be achieved by plasticization with glycerin and subsequent drying of the mixture. The storage modulus, G', of the resulting whey protein/glycerin glasses ranged from 1.14×10^4 to 3.29×10^9 dynes/cm^2 for a 10% and a 70% protein dispersion respectively (Table 1.3). The whey protein/glycerin glasses

Table 1.3 Ten second storage modulus (G') (dynes/cm^2) for whey protein isolate and gelatin in glycerin solution at 30°C[a]

Percentage protein	Storage Modulus (dynes/cm^2)	
	Whey Protein isolates	Gelatin
10	1.14×10^3	6.65×10^4
20	1.78×10^6	3.33×10^5
29	–	1.87×10^6
30	2.75×10^7	–
39	–	1.33×10^7
40	1.29×10^8	–
50	1.35×10^9	7.46×10^6
60	1.91×10^9	–
61	–	1.33×10^9
67	–	2.36×10^9
70	3.29×10^9	–

[a]Source: Yannas and Tobolsky (1966).

were stronger than gelatin/glycerin glasses at equivalent concentrations (Yannas and Tobolsky, 1966) and at all protein concentrations examined except the 10% protein dispersions (Table 1.3). The potential of forming glassy to elastic films from whey proteins and β-lactoglobulin is being studied further.

1.3 Foaming properties

Proteins are of widespread industrial importance in the food industry because of their unique characteristic of imparting desirable textural attributes to foods such as occurs after air incorporation. Air incorporation is evident in several foods including breads, cakes, whipped toppings, ice cream and meringues. Foams have been described as thermodynamically unstable colloidal systems in which gas is maintained as a distinct dispersed phase in a liquid matrix (German and Phillips, 1989). The correctness of this statement when considering food foams is debatable, yet this statement describes an initial state of foams before cooking, baking or freezing.

Our understanding of the contribution of proteins to the foaming process is often limited to empirical observation (Kinsella and Phillips, 1989). If compared with the knowledge of the biochemical functions of proteins, the information regarding structure/function relationships of proteins in foams is lacking (Kinsella and Phillips, 1989). This lack of information is, in part, caused by the lack of understanding of processes leading to protein folding and the complexity of food systems. When investigating the structure/function relationship of a functional property such as foaming, one must consider the molecular changes occurring during processing as well as the structural properties of the native protein (Li-Chan and Nakai, 1989).

Some progress has been made towards understanding protein adsorption at an air–water interface, foam formation and foam stability – especially through efforts to develop standardized methods for studying foaming properties (Phillips *et al.*, 1990a) but further research is needed to determine the three-dimensional structures for many of the food proteins involved in foaming, and to relate that structural information to the observed foaming properties (Kinsella and Phillips, 1989).

1.3.1 Interfacial adsorption of proteins

For a protein to be a successful foaming agent, it must be able to stabilize the new surface area continuously being created during foaming (MacRitchie, 1978). The migration of proteins to the interface is energetically favorable because some of the conformational energy and some of the energy of hydration of the protein is lost at the interface (Phillips, 1981).

The first stage in protein adsorption at the air–water interface is diffusion controlled, i.e. there is free movement of protein to the interface (Graham and Phillips, 1976; MacRitchie, 1978; Phillips, 1981). The next phase is marked by interfacial 'crowding'. The surface pressure increases as more protein interacts with the interface. An electrostatic barrier develops on the water side of the interface. The electrostatic barrier develops because the charged groups of the proteins orient preferentially on the aqueous side of the interface. Adequate kinetic energy is needed to overcome the electrical barrier and to compress the molecules already at the interface to allow the adsorption of additional protein (MacRitchie, 1978). Once adsorbed at an interface, protein molecules interact with the interface further to obtain a state of lowest free energy. Compact, globular proteins form a more condensed, tightly packed interfacial film than random-coil or denatured proteins (MacRitchie, 1978; Waniska and Kinsella, 1985). More protein can occupy the interface when the protein conformation is globular and does not denature readily at the interface, e.g. lysozyme (Graham and Phillips, 1976; Waniska and Kinsella, 1985).

1.3.2 Electrostatic interactions

Traditionally, the role of electrostatic interactions has been that the net negative charge on the outer film lamella helps stabilize foams via net repulsion of adjacent films; however, electrostatic attractions within the film may also be important (Kinsella and Phillips, 1989).

1.3.2.1 pH effects. The rearrangement and adsorption of β-lactoglobulin (β-Lg) during film formation at several pH levels was studied by Waniska and Kinsella (1985). The net charge on β-Lg affected its surface active properties. The optimum surface pressure occurred at pH 4.9, slightly below the isoelectric point (pI) of β-Lg (pI = 5.25). In general, the rate of interfacial protein adsorption increases near the pI of a protein when the protein remains soluble. This is because the proteins have decreased electrostatic repulsion at the interface and because more compact protein molecules can pack to a greater extent into the interfacial film (Waniska and Kinsella, 1985).

Phillips *et al.* (1990b) correlated β-Lg foaming data with interfacial data from Waniska and Kinsella (1985). They found the work of compression and rate of adsorption for β-Lg to be highly correlated with foam stability, overrun, maximum overrun and overrun development, and increased in a linear fashion (Table 1.4), whereas the rate of rearrangement and the average area cleared were not correlated to the foaming properties of β-Lg. These results may reflect those interfacial properties that are most important during the whipping process (Phillips *et al.*, 1990b). Whipping a protein imparts a certain amount of surface energy and causes the unfolding of proteins. As new air–water interfacial area is formed, the protein must quickly adsorb at and coat the newly formed surface in order to stabilize the entrapped air (Kinsella and Phillips, 1989). The rate of rearrangement and the average area cleared may not be rate-limiting steps during the whipping process compared with the passive adsorption that occurs at an unstirred interface (Table 1.4).

Table 1.4 Correlation coefficients for relationships between interfacial adsorption behavior of β-lactoglobulin with the following foaming properties[a,b]

Film parameters	Foaming properties expressed as correlation coefficient (r)			
	Foam stability	Overrun (15 min)	Overrun (maximum)	Δoverrun/Δmin
Work of compression	1.000	0.971	0.951	0.990
Adsorption	0.856	0.957	0.975	0.922
Rearrangement	0.636	0.425	0.358	0.514
Average area cleared	0.579	0.761	0.807	0.692

[a]Source: Phillips *et al.* (1990b).
[b]The tabulated correlations are Pearson coefficients of correlation and relate the degree of linearity between the variables; work of compression = work required to clear an area at the interface for insertion of a protein residue; adsorption = constant relating the relative rate of adsorption of molecules to the interface; rearrangement - constant relating the relative rate of rearrangement of molecules in the interface; average area cleared = estimate of the area of the interface that must be cleared for penetration of the molecule; foam stability = time required for 50% of the liquid to drain from a foam; overrun = the measure of foam volume; Δoverrun/Δmin = change in overrun over time as a foam is whipped.

Phillips *et al.* (1990b) concluded that the overrun obtained by whipping whey protein isolate (WPI) solution was significantly affected ($p<0.05$) by changing the pH (Phillips *et al.*, 1990b). They observed the highest overrun values for WPI were at pH 5.0 (Figure 1.4). This parallels the optimum surface pressure values obtained at pH 5.0 by Waniska and Kinsella (1985). These data provide further evidence that interfacial film formation is optimized by a balance of charge on the protein molecule at pH values near the pI. The highest foam stability values for WPI (i.e. time for 50% of the foam to collapse equaled 60.8 min) were also obtained at pH

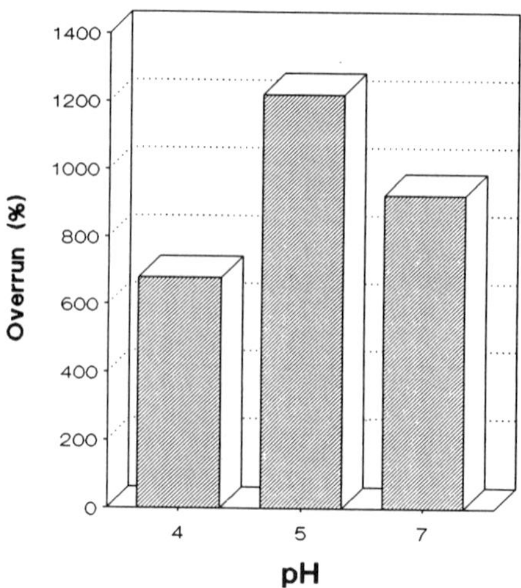

Figure 1.4 Effects of pH on the overrun of whey protein isolate (from Phillips *et al.*, 1990b).

5.0. Phillips *et al.* (1990b) remarked that pH 5.0 WPI foams held more air and were more stable than those formed at pH 7.0 or pH 4.0. These researchers also observed that heat treatment had little effect on foam stability at pH 4.0 or 7.0 but, as pH 5.0, heat treatment at 80°C for 10 min improved stability by 65%.

1.3.2.2 Electrostatic attraction. In heterogenous protein systems, such as egg white, the constituent proteins have different isoelectric points and carry different net charges. In egg white, electrostatic interactions contribute to excellent foaming and heat stability characteristics (Johnson and Zabik, 1981). At the natural pH of fresh egg white (pH 7–8), the basic protein lysozyme (pI 10.7) is positively charged and can interact electrostatically with negatively charged proteins (Phillips *et al.*, 1989a). Poole *et al.* (1984) demonstrated that the addition of low concentrations (0.01–0.1% w/v) of clupeine (pI 12) and lysozyme to solutions of acidic proteins (0.50%, pH 8) improved their foam volume and stability by as much as 260% and 600% respectively. They suggested that electrostatic interactions between basic and acidic proteins in the film enhanced foam volume an stability. The addition of sucrose together with lysozyme and clupeine enhanced the foam volume and stability by as much as 420% and 700%, respectively. They suggested that sucrose decreased the hydration of the proteins thereby facilitating their movement to the interface (Poole *et al.*, 1984).

Phillips *et al.* (1989b) demonstrated that lysozyme improved the foaming properties of β-Lg and whey protein isolate (WPI) (Figure 1.5). The stability of β-Lg and WPI foams were increased by 124% and 114%, respectively, by adding

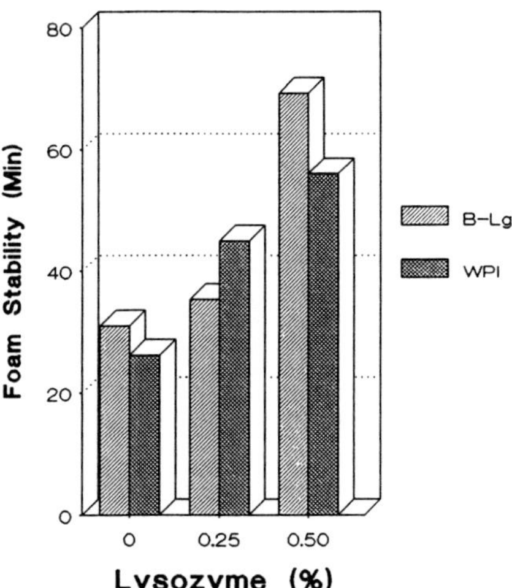

Figure 1.5 Effects of lysozyme on the foam stability of β-lactoglobulin and whey protein isolate (pH 8.0) (from Phillips *et al.*, 1989b).

0.5% lysozyme. The resultant foams were, in many ways, superior to egg-white foams. Lysozyme and clupeine also improved the heat stability of β-Lg and WPI foams. Basic proteins such as lysozyme may alter the structure of flexible acidic proteins at a pH between the pIs of the two proteins (Poole *et al.*, 1987). This interaction may lead to the formation of an elastic film, which is disrupted less rapidly during heating. Molecular size appears to be an important variable in electrostatic interactions. Poole *et al.* (1984) reported that the basic amino acid arginine did not enhance foaming, and that clupeine lost its capacity to enhance ovalbumin foams after 18.4% hydrolysis (Poole *et al.*, 1987). These studies suggest that a minimum size for the basic protein may be required for electro-static interaction and enhancement of foaming properties.

Similar to the results with β-Lg and lysozyme, it was possible to provide conditions for electrostatic attraction that improved foaming by mixing succin-ylated (negative charge added) and native β-Lg (Phillips and Kinsella, 1991b). The electrostatic interactions caused by the addition of 100% succinylated β-Lg (0.5 g/100 ml) to a 2.5% solution of native β-Lg at pH 4.0 improved overrun (47%) and foam stability (61%). The use of mixtures of proteins carry-ing opposite charges may prove very useful for stabilizing food foams once enough is known about the electrostatic interactions involved for proper control.

1.3.2.3 Electrostatic repulsion.   Interactions such as electrostatic repulsion can have a pronounced effect on foaming. Phillips and Kinsella (1991b) studied the effects of increasing the negative charge on β-Lg via succinylation. They observed a 45 and 55% reduction in foam stability for 50 and 100% succin-ylation, respectively (Figure 1.6). This was attributed to the increase in negative charge on the surface of the β-Lg molecule following succinylation which resulted in electrostatic repulsion. This repulsion hindered film and foam forma-tion (Phillips and Kinsella, 1991b).

1.3.3 Disulfide bonds

Disulfide bonds reduce the flexibility of a protein. Understanding the free thiol–disulfide interchange is of extreme importance in manipulating functional properties. In the case of soy protein, disulfide bonds not only limit molecular flexibility but also restrict foaming. German *et al.* (1985) established that the molecular alterations induced by reducing intersubunit disulfide bonds greatly improved film formation, foaming and foam stability. Cleavage of the disulfide bonds of whey proteins (containing mostly β-Lg) improved foaming, e.g. a 3.5-fold increase in overrun was observed upon cleaving all the disulfide bonds in whey protein isolate. Doi *et al.* (1989) studied the relationship between disul-fide bond formation during foaming and the foaming properties of ovalbumin. They concluded that the essential factor for formation of stable ovalbumin foams was not the formation of intermolecular disulfide bonds since disrupting these disulfide bonds did not affect foam stability. They hypothesized that the network

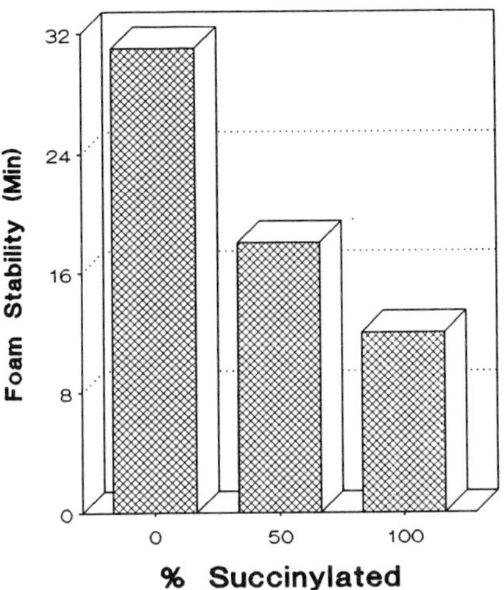

Figure 1.6 The effects of the extent of succinylation (0, 50.0 and 100.0%) on the foam stability of β-lactoglobulin at pH 7.0 (with 95% confidence intervals). The succinylated β-lactoglobulin (2.5g/ 100 ml) was whipped for 15 min and foam stability was measured (from Phillips and Kinsella, 1991).

formation by other noncovalent interactions was the main contributor to foam formation and that polymerization was just a side-effect. Li-Chan and Nakai (1989) suggested that the critical role of disulfide bonds was to stabilize protein structure, restricting the unfolding of the molecule and preventing complete exposure of buried hydrophobic regions. Phillips (1992) has shown that disulfide bond formation at an air–water interface can dramatically improve foam stability.

1.3.4 Hydrophobic effects

Hydrophobic interactions are the main driving force responsible for the structure of proteins in solution. The contribution of hydrophobic effects to the functional properties of foods has received considerable attention in recent years (German and Phillips, 1989; Phillips and Kinsella, 1991; Li-Chan and Nakai, 1989). Li-Chan and Nakai developed equations that incorporate protein hydrophobicity as a significant variable to predict functionality. This endeavor is referred to as the 'quantitative structure-activity relationship', QSAR (Li-Chan and Nakai, 1989). The importance of surface hydrophobicity has been stressed instead of 'total' hydrophobicity which is exemplified by Bigelow's hydrophobicity parameter computed as the sum of the side chain hydrophobicities (Li-Chan and Nakai, 1989). However, they found little correlation between surface hydrophobicity and foaming and suggested that the poor correlation was the result of the change in surface hydrophobicity upon unfolding a protein at an interface. Therefore, they postulated that a measure of surface hydrophobicity in solution may be quite different to the actual surface hydrophobicity of a protein at an interface (Li-Chan and Nakai, 1989). To improve their analysis, they developed a new technique for measuring the hydrophobicity after the protein has been suitably unfolded. The resulting 'exposed' hydrophobicity showed significant correlations with foam capacity (Li-Chan and Nakai, 1989).

1.3.4.1 Neutral salts. Neutral salts such as Na_2SO_4, NaCl and NaSCN affect the physicochemical properties and interactions between proteins either by ionic strength effects, binding to the protein charged groups or at high concentrations by altering water structure, with subsequent changes in hydrophobic effects. Salts can be effective probes for studying the effects of altering hydrophobicity on foaming.

Adding neutral salts to WPI affects the foaming properties (Phillips *et al.*, 1991b). High concentrations (1 M) of Na_2SO_4 improved foam stability by 76% compared with WPI without salt (Figure 1.7) (Phillips *et al.*, 1991b). Chloride had an intermediate effect, whereas NaSCN did not improve foam stability. Increasing the Na_2SO_4 concentration (2 M), however, improved foam stability by 127% compared with the control (Phillips *et al.*, 1991b). The stability of WPI foams to heat treatment was also enhanced by Na_2SO_4 and NaCl (1 M) by 83% and 26%, respectively, whereas NaSCN decreased heat stability by 23%. The different anions may have been exerting their effects via electrostatic interactions, i.e. by differential binding to the proteins, and by altering hydrophobic effects through restructuring of water, thus leading to an increase in protein–protein interactions (Damodaran and Kinsella, 1982). The relative effectiveness of the

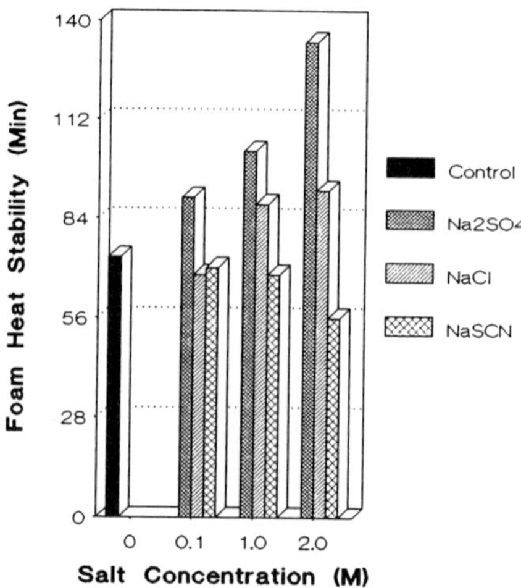

Figure 1.7 The effects of neutral salts on the foam stability of whey protein isolate (from Phillips *et al.*, 1991b).

salts at improving foam stability and foam heat stability followed the Hofmeister series (SO_4^{2-}>Cl^->SCN^-). The effect of SCN^- on heat stability may reflect its chaotropic effect, by perturbing water structure the anion may reduce hydrophobic effects and weaken the film. Conversely, SO_4^{2-} and Cl^- (structure-forming ions) may stabilize the protein against heat denaturation, thereby retarding foam collapse (Phillips *et al.*, 1991b).

1.3.4.2 Temperature effects. Hydrophobic effects are reduced as the temperature is lowered (Kinsella and Phillips, 1989). The importance of temperature on hydrophobic effects and on foam formation are demonstrated by the observed decrease in foam stability of β-Lg with a decrease in temperature (Figure 1.8). Phillips *et al.* (1991a) observed an eight-fold reduction in the foam stability of β-Lg when the temperature was reduced from 25°C to 3°C. They attributed the decrease to reduced hydrophobic effects, causing poor film formation and resulting in reduced foam stability.

1.3.5 Foam depressants

Foam depressants can lead to the loss of liquid from a foam or the complete disruption of a foam. Foams are disrupted when thin spots form and cause rupture of the films stabilizing the entrapped air (Adamson, 1982; Prins, 1988). Thin-spot formation is precipitated by one of the following mechanisms: (i) a pressure gradient in the film liquid; (ii) a surface tension gradient acting on the film; or (iii) evaporation of the liquid in the film (Prins, 1988). Pressure-gradient

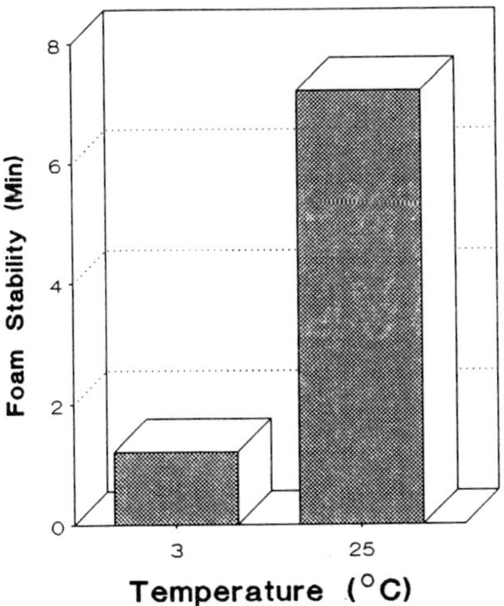

Figure 1.8 The effects of temperature during whipping on β-lactoglobulin foams (from Phillips *et al.*, 1991a).

formation can be caused by the insertion of a hydrophobic particle into a film. The hydrophobic particle must be large enough to span the film. It ruptures the film by causing locally high pressures, which leads to liquid flowing away from the particle. The process continues until the liquid breaks contact with the particle and a hole is produced (Prins, 1988).

The second type of foam depressant involves a surface-tension gradient. This can occur by displacing surfactant that is needed to stabilize the foam, or by spreading of a foam depressant onto the film surface, which causes a local movement of other particles away from the foam depressant leading ultimately to rupture (Prins, 1988). Prins described an experiment where emulsified soy oil was tested as a foam depressant of caseinate foams. He observed maximum foam depressant effect for fat droplets with 2–4 µm diameters. Outside this size range no effect was observed. Phillips *et al.* (1987, 1989a) isolated a foam depressant from milk. It contained protein as well as fat and was retained on a 100 000 molecular weight cut-off membrane. The foam depressant reduced the foam stability of egg white foams and, in high enough concentration (0.25%), completely inhibited egg white foams (Phillips *et al.*, 1987). The foam depressant, when added at 0.1% concentration to WPI and sodium caseinate, reduced overrun (Figure 1.9). (Phillips *et al.*, 1989a). This foam depressant is a major problem with whey processing because it is often present in whey and can adversely affect the foaming properties of whey.

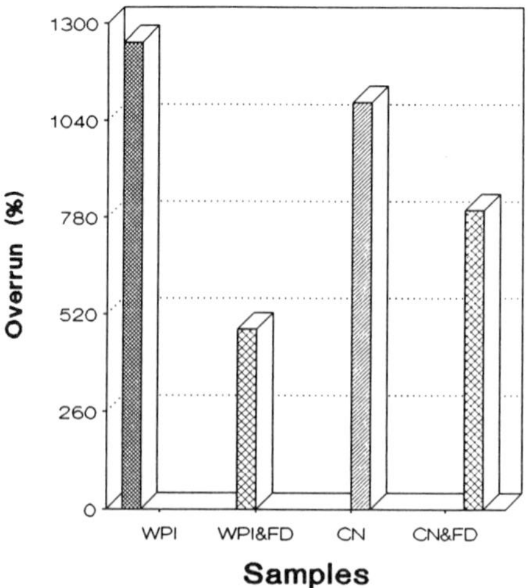

Figure 1.9 The effects of foam depressant collected from milk on the overrun of whey protein isolate and casein. WPI=whey protein isolate; WPI&FD=whey protein isolate with foam depressant added (0.1%); CN=sodium caseinate; CN&FD-sodium caseinate with foam depressant added (0.1%) (from Phillips *et al.*, 1989a).

1.3.5.1 Sucrose. Phillips *et al.* (1989b) observed that sucrose reduced the overrun and foam stability of β-Lg, although added sucrose (10%) increased the specific viscosity four-fold (Table 1.5). Sucrose had little effect on the overrun or foam stability of WPI (see Table 1.2). The major effect of sucrose was on the heat stability of the β-Lg and WPI foams. In addition, Phillips et al. (1989b) observed a four-fold increase in the heat stability of β-Lg foams with 10% sucrose and a 30% increase for WPI foams (Table 1.5). Lee and Timasheff (1981) showed that the stabilizing effect of sucrose originates from the preferential hydration of proteins in solution. The sucrose may allow formation of a better network in protein films instead of random aggregation during heating, which would disrupt the film (Poole et al., 1987).

Table 1.5 The effects of sugar (10%) on the viscosity and foaming properties of β-lactoglobulin (2.5%) and whey protein isolate (5%)[a]

		Foaming properties		
		Maximum	Foam	Foam heat
	Viscosity	overrun	stability	stability
Treatment	(cp)	(%)	(min)	(%)
β-lactoglobulin	0.0865	1583	30.9	17.8
β-lactoglobulin + 10% sucrose	0.3959	1234	25.9	90.7
Whey protein isolate	—	663	26.2	76.3
Whey protein isolate + 10% sucrose	—	626	31.1	98.6

[a]Source: Phillips *et al.* (1989b).

Proteins because of their dynamic structures and amphiphilic nature possess varying interfacial properties. This overview has provided some clues of molecular structure to function relationships. The multitude of possible interactions underscore the challenge of completely describing and controlling all the factors and interactions that control foam formation. The dynamic nature and flexibility of proteins indicate the possibility of further enhancing the foaming properties of proteins. More emphasis needs to be placed on understanding how varying processing conditions affect the flexibility and the interaction of proteins for the purpose of controlling functionality.

This work indicates the need for continuing research that attempts to interrelate the chemistry, structure, conformation and physicochemical properties of food proteins. Developing new information will expand uses for food macromolecules and extend product development.

Acknowledgements

This research was supported by grants from the National Dairy Board and Wisconsin Milk Marketing Board.

References

Adamson, A. (1982) *Physical Chemistry of Surfaces*, 4th edn, John Wiley, New York.

Atkins, P.W. (1982) *Physical Chemistry*, W.H. Freeman, New York, p. 266.

Beveridge, T., Jones, L. and Tung, M. (1984) Progel and gel formation and reversibility of gelation of whey, soy beans and albumin protein gels. *J. Agric. Food Chem.* **32**:307–13.

Catsimpoulas, N. and Meyer, E.W. (1970) Gelation phenomena of soybean globulins. I. Protein–protein interactions. *Cereal Chem.* **47**:559–70.

Clark, A.H. and Lee-Tuffnell, C.D. (1986) Gelation of globular proteins, in *Functional Properties of Food Macromolecules*, J.R. Mitchell and D.A. Ledward (eds) Elsevier, London, pp. 203–72.

Clark, A.H., Richardson, R.K., Robinson, G., Ross-Murphy, S. and Weaver, A.Q. (1982) Structure and mechanical properties of agar/BSA cogels. *Prog. Food Nutr. Sci.* **6**:149–60.

Doi, E., Kitabatake, N., Hatta, H. and Kaseki, T. (1989) Relationship of SH groups to functionality of ovalbumin, in *Food Proteins: Structure and Functional Relationships*, (eds J.E. Kinsella and W. Soucie), American Oil Chemists' Society, Champaign, Illinois.

Eldridge, J.E. and Ferry, J.D. (1954) Studies of the cross-linking process of gelatin gels. III. Dependence of melting point on concentration and molecular weight. J. Physical Chem. **58**:992–95.

Ferry, J.D. (1948) Protein gels. *Adv. Protein Chem.* **4**:1–78.

German, J., O'Neill, T. and Kinsella, J. (1985) Film forming and foaming behaviour of food proteins. *J. Am. Oil Chem. Soc.* **62**:1358.

German, J.B. and Phillips, L.G. (1989) Structures of proteins important in foaming in *Food Proteins: Structure and Functional Relationships*, (eds J. Kinsella and W. Soucie), *American Oil Chemistry Society*, Champaign, Illinois.

Graham, D.E. and Phillips, M.C. (1976) The conformations of proteins at the air–water interface and their role in stabilizing foams, in *Foams*, (ed. R.J. Akers), Academic Press, New York, p.233.

Haque, Z. and Kinsella, J. (1988) Interactions between κ-casein and β-lactoglobulin: Predominance of hydrophobic interactions in the initial stages of complex formation. *J. Dairy Res.* **55**:67–80.

Hermansson, A.M. (1979) Aggregation and denaturation involved in gel formation, in *Functionality and Protein Structure* (ed. A. Pour-El) American Chemical Society, Washington DC, pp. 82–103.

Johnson, T. and Zabik, M. (1981) Egg albumen proteins interactions in an angel food cake system. *J. Food Sci.* **46**:1231.

Kella, N.K. (1987) Effects of amides and urea on the reversible gelation of arachin. *Int. J. Biol. Macromol.* **9**:238–44.

Kella, N.K. and Kinsella, J. (1988) Structural stability of β-lactoglobulin in the presence of kosmotropic salts. *Int. J. Peptide Res.* **32**:396:405.

Kella, N., Yang, T. and Kinsella, J. (1989) Effect of disulfide bond cleavage on structural and interfacial properties of whey proteins. *J. Agric. Food Chem.* **37**:1203.

Kella, N., Nandi, P. and Rao, M. (1980) Reversible gelation of arachin. *Int. J. Peptide Protein Res.* **15**:67–72.

Kinsella, J.E. (1976) Functional properties of proteins in foods: A survey. *Crit. Rev. Food Sci. Nutr.* **7**:219–80.

Kinsella, J.E. (1982) Relationship between structure and functional properties of food proteins, in *Food Proteins*, (eds P.F. Fox and J.J. Condon), Applied Science, New York. pp. 51–103.

Kinsella, J.E. (1984a) A review. Milk proteins: physicochemical and functional properties. *CRC Crit. Rev. Food Sci. Nutr.* **21**(3):197–262.

Kinsella, J.E. (1984b) Functional properties of proteins: Thermal modification involving denaturation and gelation, in *Food Science and Technology Present Status and Future Direction*, Proceedings of the Sixth International Congress of Food Science and Technology, September 1983, Boole Press, Dublin, pp. 226–46.

Kinsella, J.E. and Phillips, L.G. (1988) Structure function relationships in food proteins: Film and foaming behavior, in *Food Proteins: Structure and Functional Relationships*. (eds J.E. Kinsella and W. Soucie), American Oil Chemistry Society, Champaign, Illinois.

Lee, J. and Timasheff, S. (1981) The stabilization of proteins by sucrose. *J. Biol. Chem.* **246**:7193.

Li-Chan, E. and Nakai, S. (1989) Effects of molecular changes (SH groups and hydrophobicity) of food proteins on their functionality, in *Food Proteins: Structure and Functional Relationships*, (eds J.E. Kinsella and W. Soucie), American Oil Chemistry Society, Champaign, Illinois, pp. 232–51.

MacRitchie, F. (1978) Proteins at interfaces. *Adv. Protein Chem.* **32**: 283–326.

Mulvihill, D.M. and Kinsella, J.E. (1987) Gelation characteristics of whey protein and β-lactoglobulin. *Food Technol.* **41**:102–11.

Mulvihill, D.M. and Kinsella, J.E. (1988) Gelation of β-lactoglobulin: Effects of sodium chloride and calcium chloride on the rheological and structural properties of gels. *J. Food Sci.* **53(1)**:231–36.

Phillips, L.G. (1992) *The Relationship Between Structural, Interfacial and Foaming Properties of β-lactoglobulin.* PhD Thesis, Cornell University, Ithaca, New York.

Phillips, L.G. and Kinsella, J.E. (1991) Effects of succinylation on β-lactoglobulin foaming properties. *J. Food Sci.* **55(6)**:1735.

Phillips, L.G., Haque, Z. and Kinsella, J.E. (1987) A method for the measurement of foam formation and stability. *J. Food Sci.* **52**:1047–49.

Phillips. L.G., Davis, M.J., Kinsella, J.E. (1989a) The effects of various milk proteins on the foaming properties of egg white. *Food Hydrocoll.* **3(3)**:163–74.

Phillips, L.G., Yang, S.T., Schulman, W. and Kinsella, J.E. (1989b) Effects of lysozyme, clupeine and sucrose on the foaming properties of whey protein isolate and β-lactoglobulin. *J. Food Sci.* **54(3)**:743.

Phillips, L.G., German, J.B., O'Snail, T.E. Foegeding, E.A., Harwalkar, V.R., Kilara, A., Lewis, B.A., Mangino, M.E., Morr, C.V., Regenstein, J.M., Smith, D.M., and Kinsella, J.E. (1990a) Development of a standardized method for the measurement of the foaming properties of food proteins: a collaborative study. *J. Food Sci.* **55(5)**:1441.

Phillips, L.G., Schulman, W. and Kinsella, J.E. (1990b) pH and heat treatment effects on foaming of whey protein isolate. *J. Food Sci.* **55(4)**:1116.

Phillips, L.G., Hawks, S.E. and Kinsella, J.E. (1991a) *The effects of shear and temperature on the foaming properties of β-lactoglobulin.* Presentation at the American Chemical Society Meeting, New York, 25–30 August.

Phillips, L.G., Yang, S.T. and Kinsella, J.E. (1991b) Effects of neutral salts on the stability of whey protein isolate foams. *J. Food Sci.* **55(2)**:588.

Phillips, M.C. (1981) Protein conformation at liquid interfaces and its role in stabilizing emulsions and foams. *Food Technol.* **35**:50–57.

Poole, S., West, S., and Fry, J. (1987) Effects of basic proteins on the denaturation and heat gelation of acidic proteins. *Food Hydrocoll.* **1(4)**:301.

Poole, S., West, S., and Walters, C. (1984) Protein–protein interactions: Their importance in the foaming of heterogenous protein systems. *J. Sci. Food Agric.* **35**:701.

Prins, A. (1988) Principles of foam stability, in *Advances in Food Emulsions and Foams*, (eds E. Dickinson and G. Stainsby), Elsevier Applied Science, New York.

Rector, D.J. (1992) *Polymer Properties of Whey Proteins: Gelation, Viscous and Glassy Behavior.* MS Thesis, Cornell University, Ithaca, New York.

Rector, D., Kella, N. and Kinsella, J. (1991) Reversible gelation of whey proteins: Melting, thermodynamics and viscoelastic behavior. *J. Texture Studies* **20**: 457–71.

Rodriguez, F. (1982) *Principles of Polymer Systems*, 2nd edn., Hemisphere Publishing, New York, pp. 17–18.

te Nijenuis, K. (1981a) Investigation into the aging process in gels of gelatin/water systems by measurement of their dynamic moduli. Part I – Phenomenology. *Colloid Polymer Sci.* **259**:522–35.

te Nijenuis, K. (1981b) Investigation into the aging process in gels of gelatin/water systems by measurement of their dynamic moduli. Part II – Mechanism of the aging process. *Colloid Polymer Sci.* **259**:522–35.

Waniska, R.D. and Kinsella, J.E. (1985) Surface Properties of β-Lactoglobulin: Adsorption and Rearrangement during Film Formation. *J. Agric. Food Chem.* **33**:1143.

Xiong, Y. and Kinsella, J. (1990) Mechanism of urea-induced whey protein gelation. *J. Agric. Food Chem.* **38**:1888–91.

Yanas, J. and Tobolsky, A. (1966) Transitions in gelatin-non-aqueous-diluent systems. *J. Macromol. Chem.* **1(4)**:723–37.

Young, R.J. (1983) *Introduction to Polymers.* Chapman and Hall, New York, pp.3–4.

2　Microstructure of protein gels related to functionality

A.-M. HERMANSSON

Abstract

Gel structures are responsible for many physical properties such as water-holding and rheological properties. Gels can schematically be divided into fine-stranded and aggregated or particulate gels. The gel structure of transparent fine-stranded gels can only be revealed by high-resolution transmission electron microscopy, whereas opaque aggregated gels can also be studied by scanning electron microscopy at lower magnification and sometimes even by light microscopy. Image analysis can be used to quantify network parameters. The water-holding properties of gel networks are determined mainly by the pore-size distribution and a more open structure gives rise to poorer water-holding than a dense network structure. Rheological properties can either be measured by small-scale deformation tests where the structure remains intact or by large deformation tests resulting in fracture. Complementary information can be obtained about gel structures from a combination of small- and large-scale deformation properties.

2.1 Introduction

A specific food product can be made using a variety of formulations and processes, which differ in their demand on the functional ingredients. Proteins with quite different functional characteristics may give the same contribution to overall product quality, even if achieved in different ways. Thus, soy protein isolate, sodium caseinate and blood plasma or whey proteins may all give a positive contribution to the binding properties of comminuted meat products. Soy protein isolate achieves this mainly by means of its swelling or gelling properties; caseinate achieves this by its emulsifying properties and high viscosity, despite the fact that it lacks the ability to form heat-induced gels; and low viscosity blood plasma and whey protein solutions achieve this by forming firm gels on heating. Proteins, as well as starch and other polysaccharides, are commonly added to foods to enhance such characteristics as texture, fat- and water-holding properties, and foam and emulsion stability. A better understanding of these functional properties may be obtained from a knowledge of the food microstructure. The purpose of food processes such as the production of bread, cheese and sausages is to create a structure that gives the food item its characteristic properties.

Several proteins have the ability to form gels on heating. This ability is important in most half-solid foods such as milk, as well as in meat, fish, egg and cereal products. The type of network structure formed contributes to the characteristic texture of many food products and acts as a matrix, holding water as well as fat and other components. The potential of using protein gel structures to bind water and mimic texture in low-fat products is a new and challenging area. Even if it is well-known that the gel microstructure relates strongly to functional properties, far too little is known about gel structures in general, especially at the colloidal and supramolecular level, and about structure–function relationships.

Many proteins have the ability to form different types of network structures depending on factors such as temperature, pH and the presence of salts and specific ions (Clark and Lee-Tuffnell, 1986; Clark and Ross-Murphy, 1987; Hermansson, 1988; Ziegler and Foegeding, 1990). Figure 2.1 shows examples of different types of structures common for biopolymer gels. Such gels can be divided roughly into fine-stranded and coarse aggregate gels. Fine-stranded gels shown in Figures 2.1a and 2.1b are formed by an ordered association of molecules, and the dimensions are often so small that these gels are transparent. Proteins can also form particulate gels, illustrated in Figures 2.1c and 2.1d, that are not transparent. These latter types of gels are common in, for example, milk and egg products. The size of the aggregates can vary considerably depending on the type of protein(s) and environmental conditions, but the aggregates are relatively uniform in size within a particular gel.

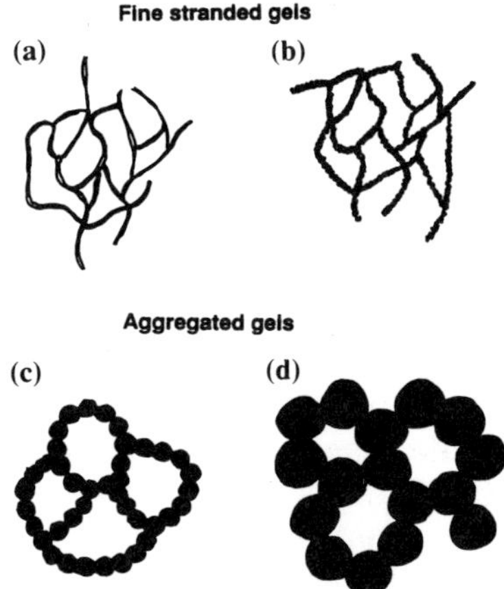

Figure 2.1 Diagrams of different types of gel networks: (a) a fine-stranded gel where the strands are made up of macromolecules aligned into supramolecular assemblies; (b) a fine-stranded network of globular proteins; (c) and (d) illustrate particulate gels.

This chapter illustrates how information about the microstructure can be obtained by electron microscopy on several dimensional levels including molecules, supramolecular assemblies and gel networks. It will also demonstrate how gel structures relate to water-holding and rheological properties and how image analysis can be used to quantify network parameters.

2.2 Measuring techniques

2.2.1 Microstructure

The microstructure of protein gels can be characterized by electron microscopy. For complete characterization several techniques are necessary, both to obtain details at several orders of magnitude and to eliminate possible artifacts (Hermansson and Langton, 1993). Gels often contain more than 90% water: to observe the gel structure under a microscope this water must be solidified or removed during preparation without changing the original gel structure. This is a difficult step and several different methods of sample preparation should be used to provide corroborative data. The water can be solidified by rapid freezing, without formation of detectable ice crystals. However, the formation of artificial networks induced by ice-crystal formation is a common artifact. An alternative method is to chemically strengthen the gel network structure and to replace the water with solvents that can be critical-point-dried for scanning electron microscopy (SEM) or further polymerized and thin-sectioned for transmission electron microscopy (TEM).

Table 2.1 shows the microscopic techniques used for the visualization/ characterization of gel structures. SEM is suitable for nontransparent gels, such as the coarse aggregated structures, but TEM is necessary for visualization of transparent gels for which a higher resolution is necessary. For multiphase systems and mixed gels such as starch gels, light microscopy (LM) is a valuable aid to study the distribution of the various phases. The most common preparation technique for SEM is by chemical fixation, dehydration and critical-point-drying, but cryo-techniques can also be used for coarsely aggregated systems.

Structural details of fine-standard networks and aggregated gels can be observed using TEM. Preparation techniques generally involve chemical fixation, dehydration, embedding and thin-sectioning. Cryo-techniques offer alternative means of sample preparation. Freeze-etching is by far the most common cryo-technique used for preparation of gels; here a very small sample is rapidly frozen, fractured and etched, whereafter the surface is shadowed and replicated.

To observe gel structure at the molecular level, negative staining or mica techniques can be used in which the molecules or supramolecular assemblies are spread on a particular surface. In negative staining the surface or surrounding media, but not the molecules of interest, are stained. Mica techniques involve the use of freshly cleaved mica, which has a very smooth surface; adsorbed molecules are rotary shadowed and then replicated.

Table 2.1 Analysis of gel microstructure

Gel type	Microscopic technique
Coarse-aggregated gels	SEM, TEM
Fine-stranded gels	(SEM), TEM
Composite gels	SEM, LM, TEM
Gel components	*TEM*
Aggregates	Freeze-etching
Mixed domains	Thin sectioning
Superstrands	Negative staining
Strands	High-resolution metal shadowing
Molecules	Immuno-techniques

All these preparation techniques have recently been reviewed for the study of food biopolymers and examples of some protein gels will be given in this chapter (Hermansson and Langton, 1993).

2.2.2 Rheology

Rheological measurements can be divided into those that induce small or large deformations. These two types of tests give complementary information but need not be correlated. Small deformation tests probe viscoelastic parameters. These parameters are commonly derived by dynamic oscillatory testing in the linear viscoelastic region of the test material. Dynamic measurements allow gelation to be monitored easily since the induced deformations are usually so small that their effect on structure is negligible (Ross-Murphy, 1984). Two independent parameters are obtained from dynamic measurements: the storage modulus (G') describes the amount of energy that is stored elastically in the structure and the loss modulus (G'') is a measure of the energy loss or the viscous response. The phase angle, δ, is a measure of how much the stress and strain are out of phase with each other, and

$$\tan \delta = G''/G'$$

The phase angle is $0°$ for a completely elastic material and $90°$ for a purely viscous fluid.

Large deformation tests measure stress, strain and failure properties of a given material. Large deformation tests of biopolymer gels can be performed both in compression and in tension. Compression tests are more common but tensile tests give a clearer picture of the stresses in the sample (Luyten, 1988). The total stress during compression is the sum of both tensile and shear stresses, whereas the shear stresses in tensile testing are negligible. Other advantages of tensile testing are that the energy incurred is used only for deformation and not for friction, the start of fracture can be determined more precisely and it is possible to study the notch sensitivity of the material (Luyten, 1988; Stading and Hermansson, 1991).

As will be demonstrated, different types of information can be obtained from viscoelastic measurements at small deformations, and from fracture properties at large deformations.

2.2.3 Fat- and water-holding properties

The ability of a gel structure to hold water, as well as fat and other components, is important for many food products. Common methods for measuring water-holding properties are based on the application of an external force and subsequent measurement of the expressed water (Hermansson, 1986). The amount of water released will depend on experimental conditions, and these must be carefully considered for each type of material to be tested. An important factor is to minimize the degree of deformation during the measurement, since large deformations can result in structural changes in the material that alter the water-holding properties.

Water in excess of hydration water is entrapped in 'microheterogeneities' or pores. In a food product a broad range of pore sizes is to be expected where certain ranges will be critical for the required water-holding properties. Loosely bound water in larger pores may be important for sensory properties such as juiciness and flavour release, but will easily be lost during processing and result in undesirable weight loss. The finer the pore-size distribution, the better will be the water-holding properties. Capillary forces are strong in gels with pore sizes in the range 1–100 nm, i.e. the water is firmly held. Water lost from pores in this size range is more likely the result of syneresis due to spontaneous shrinkage of the structure than the result of an external force. Differences in the pore-size distribution in this range may be important for the properties of a food product when there is competition for water between different structures in the same product. Osmotic forces contribute to the water-holding capacity of gels. The relative importance of capillary and osmotic forces in, for example, polysaccharide gels with charged groups, has not been investigated. Pore sizes in the range 0.1–10 μm are of importance for the water-holding properties of coarsely aggregated gels. In a complex food product, various structural components may contribute to discrete ranges of pore sizes from 0.001–100 μm or larger, and ranges of importance for the water-binding properties of various products need to be further studied.

2.2 Microstructure of gels and gel formation

As already discussed, gel formation can be elucidated using electron microscopy, but it is necessary to characterize the structure on several scales or dimensional levels with a combination of preparation techniques. At high resolution, information can be obtained at the molecular level, i.e. how macromolecules associate into supramolecular assemblies. Supramolecular assemblies constitute the strands in physical gel networks and are either in the form of aligned molecules or of aggregates of globular molecules linked together. The packing of molecules, as well as the flexibility of the strands, have bearing on the network properties.

The next structural level includes two- or three-dimensional network structures of gels or surface layers, as well as initial phase-separation phenomena. The structures formed on this level determine many important functional properties such as water-holding, diffusion and rheological properties. The range at this level is quite wide. The pore size can vary from 5 nm to more than 1000 nm between a fine-stranded and a coarsely aggregated gel network.

Many food systems are complex and composed of proteins that are incompatible with each other and with other phases. For such systems it is important to determine the distribution of phases. Combinations of techniques for electron and light microscopy are often needed for the evaluation of multiphase systems.

The protein myosin is suitable for a case study of structure formation of food biopolymers because of its characteristic shape, which makes it easy to identify as an individual molecule as well as in early stages of association prior to gel formation. Figure 2.2 shows myosin molecules from bovine semimembranous muscle spread on a mica surface, rotary shadowed and replicated. The molecules are characterized by two globular or pear shaped heads and a tail of about 155 nm in length (Elliot and Offer, 1978; Walker *et al.*, 1985). Myosin is one of the most important myofibrillar proteins. In the thick filaments of muscle the myosin protein molecules are packed so that the tails form the backbone and the heads swing out from the surface of the filaments. The myofibrils dissolve at high salt concentration, which makes it possible to extract and fractionate pure myosin protein. Studies of myosin solutions at different salt concentrations have shown that markedly different types of supramolecular assemblies can be formed *in*

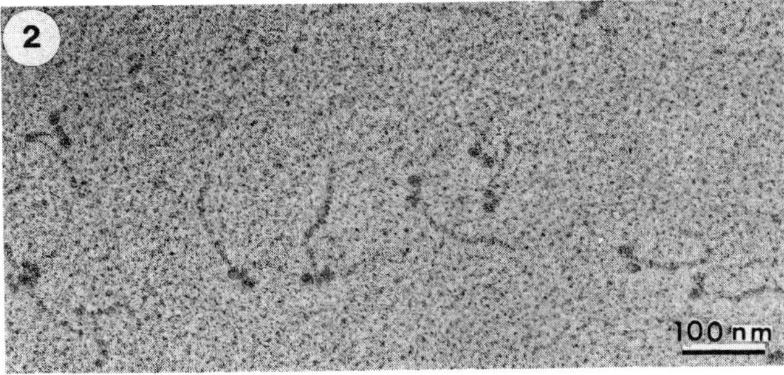

Figure 2.2 TEM-micrograph of myosin molecules from bovine *M. semimembranous* spread on mica, rotary shadowed and replicated (from Hermansson, 1988).

vitro (Hermansson, 1988; Hermansson and Langton, 1988; Yamamoto *et al.*, 1988). Rotary shadowing of dilute solutions at 0.6 M KCl heated in the range 35–45°C revealed that myosin associated in an ordered manner that was considerably different from the way myosin associated into filaments at 0.25 M KCl. Common to these associations were head-to-head interactions. Figure 2.3 shows a type of assembly with periodic association of heads and tails, as well as association into spherical micelle-like structures. In addition to these ordered supramolecular assemblies, both individual molecules and larger unordered aggregates were seen.

If the salt concentration was lowered by dialysis or dilution prior to heating, the myosin solution became turbid. Electron micrographs suggest that this is due to the formation of synthetic filaments similar in nature to those of the native thick filaments (Huxley, 1963; Kaminer and Bell, 1966; Hermansson and Langton, 1988; Yamamoto *et al.*, 1988). Figure 2.4 shows filaments formed by dialysis against pH 5.5 and 0.25 M KCl and prepared for TEM by negative staining. The filament has a backbone width of approximately 25 nm surrounded by a fringe of globular material, the myosin heads. The filaments formed in aqueous solution at low ionic strength (0.25 M KCl) are the origin of the strands of gels formed upon heating. Thus, conditions necessary for formation of this type of network are predetermined before heat treatment and gel formation. Figure 2.5 shows filaments after heating a dilute myosin dispersion at 60°C for 30 min and prepared by negative staining for TEM. The heads on the surface of the filaments melted, i.e. denatured, but the shape of the filaments remained intact after heat treatment. Figure 2.6 shows a SEM micrograph of the same myosin gel structure. Both Figures 2.5 and 2.6 show that the strands in the gel are made up of heat-denatured filaments. A comparison of the two micrographs demonstrates that the same type of information has been obtained from two entirely different preparation techniques for electron microscopy, one where a

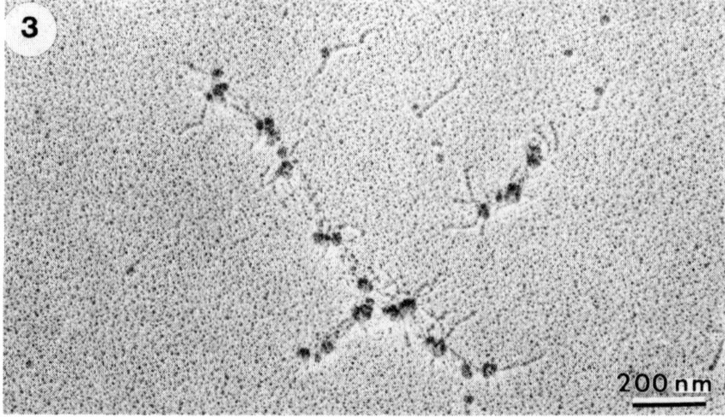

Figure 2.3 TEM-micrograph of aggregated myosin at pH 6.0 and 0.6 M KCl after heat-treatment to 40°C. The sample was prepared by the mica technique as in Figure 2.2.

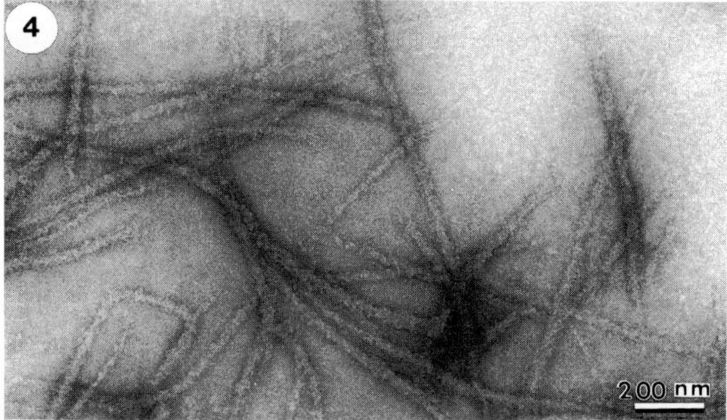

Figure 2.4 Filaments of myosin formed by dialysis against pH 5.5 and 0.25 M KCl. The filaments were prepared for TEM by negative staining (from Hermansson and Langton, 1988).

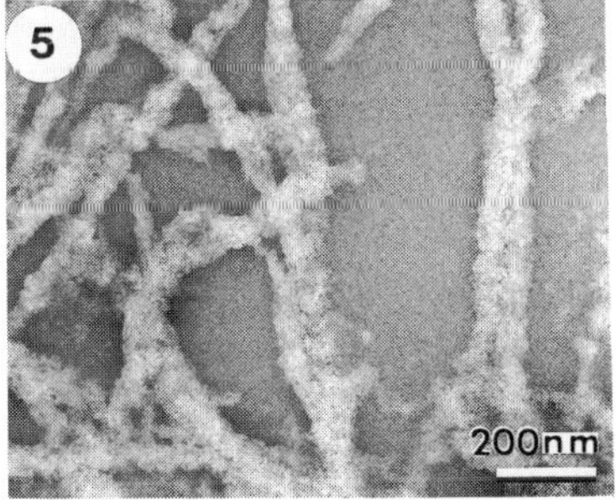

Figure 2.5 TEM micrograph of a synthetic filament of myosin at pH 5.5 and 0.25 M KCl after heat-treatment of a dilute solution at 60°C, and negatively stained (from Hermansson and Langton, 1988).

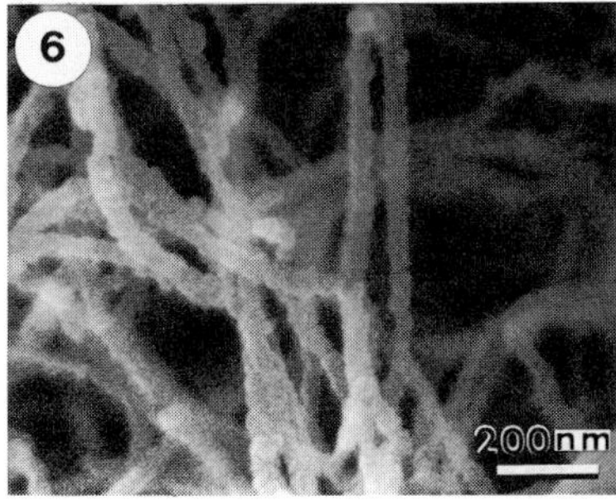

Figure 2.6 SEM micrograph showing details of the fine-stranded gel structure formed at pH 5.5 and 0.25 M KCl after heat treatment at 60°C (from Hermansson and Langton, 1988).

monolayer of myosin from dilute solution has been spread on a surface and negatively stained and one where the gel structure has been chemically fixed, dehydrated and critical-point-dried.

A quite different type of gel forms on heat treatment of myosin solutions at 0.6 M KCl and pH 5.5–6.0. As already discussed, ordered supramolecular assemblies were found after heat treatment at 35–45°C. However, these ordered structures had little influence on the final gel structure, which was composed of unordered aggregates. The order seems to collapse completely above the denaturation temperature (57–59°C) of the myosin heads at the higher salt concentration. However, aggregated gels formed already below this temperature and precursors of the ordered domains can be seen in some aggregates. Figure 2.7 shows a thin section of a myosin gel formed at 45°C. The arrow points at a region where the ordered arrangements of heads and tails still can be seen. These structures are finer in the thin section than in the metal-shadowed replica shown in Figure 2.2 but relevant comparisons can still be made. In gels formed at 60°C and above, a complete phase-separation and aggregation of proteins took place. Figure 2.8 shows part of an aggregate formed at 60°C from a myosin solution at pH 6.0 and 0.6 M KCl, in which the structural identity of the myosin molecules was lost and an unordered aggregate formed. The difference in results obtained at 35–45°C and at 60°C and above illustrate a danger of extrapolating results obtained under one set of conditions to those under another set of conditions. The ordered associations of myosin molecules at 35°C has little to do with the coarsely aggregated gel structure formed at higher temperatures, which is of more relevance to food systems. The exact gelation mechanism is currently under debate (Yamamoto, 1990). The enormous difference in the final gel structures of the filamentous gels formed at 0.25 M KCl and of the aggregated gels formed at 0.6 M KCl are clearly illustrated in the SEM micrographs shown in Figures 2.9 and 2.10. As discussed previously, the strands of the gel formed at 0.25 M KCl are composed of heat-denatured filaments (Figure 2.9) whereas the strands of the

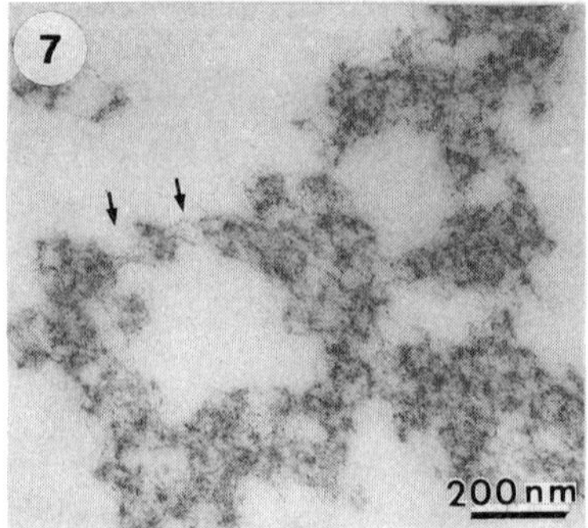

Figure 2.7 TEM-micrograph of a thin-section of an aggregate-type myosin gel formed at pH 6.0 and 0.6 M KCl after heat-treatment at 45°C.

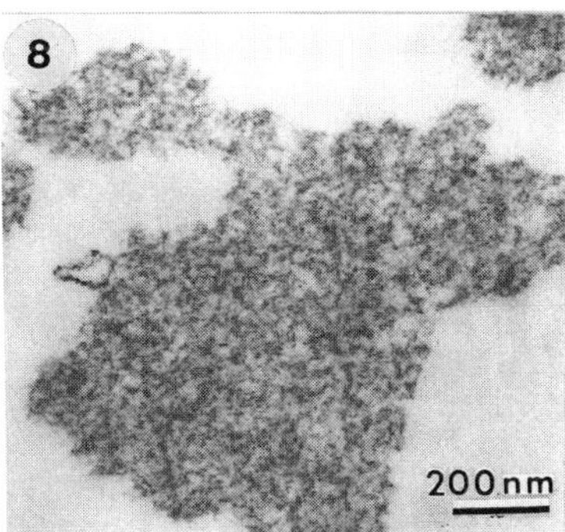

Figure 2.8 TEM-micrograph of a thin section of an aggregate-type myosin gel formed at pH 6.0 and 0.6 M KCl after heat-treatment at 60°C.

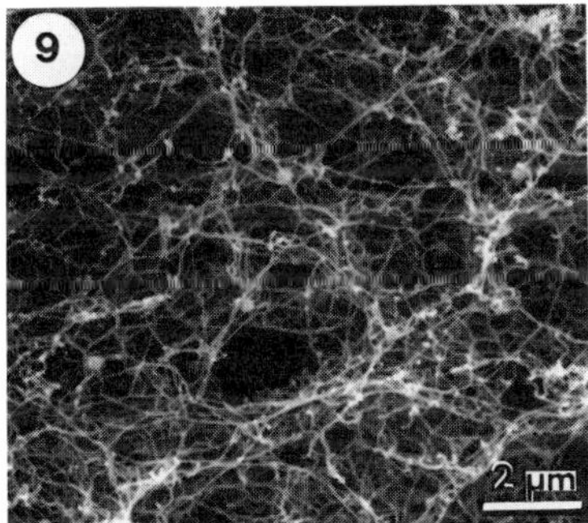

Figure 2.9 SEM micrograph of a fine-stranded myosin gel at pH 6.0 and 0.6 M KCl after heat treatment at 60°C (from Hermansson *et al.*, 1986).

Figure 2.10 SEM micrograph of an aggregated myosin gel at pH 5.5 and 0.25 M KCl after treatment at 60°C (from Hermansson *et al.*, 1986).

gel formed at 0.6 M KCl are composed of small aggregates linked together (Figure 2.10).

2.3 Gel microstructure and water-holding properties

Pore-size distribution determines the water-holding capacity of a gel. In general, the more open the structure the lower the water-holding capacity. Coarsely aggregated gels tend to have a lower water-holding capacity than fine-stranded gels. A change in the structure of a protein gel can be induced by factors such as heating temperature, pH or the addition of salt. Figures 2.11 and 2.12 show the structure of whey protein gels heated to 75°C and 95°C, respectively. They show the development of a coarser structure with larger aggregates and voids with temperature (Hermansson, 1986). Figure 2.13 shows the moisture loss of whey and blood plasma protein gels as a function of temperature. The change in gel

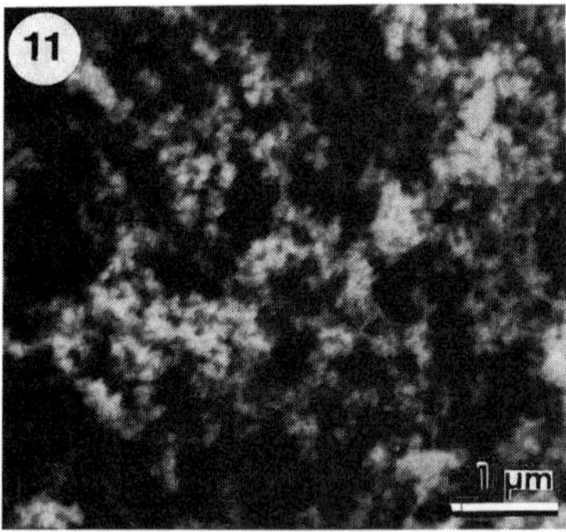

Figure 2.11 SEM micrograph of whey protein gels heated to 75°C (from Hermansson, 1986).

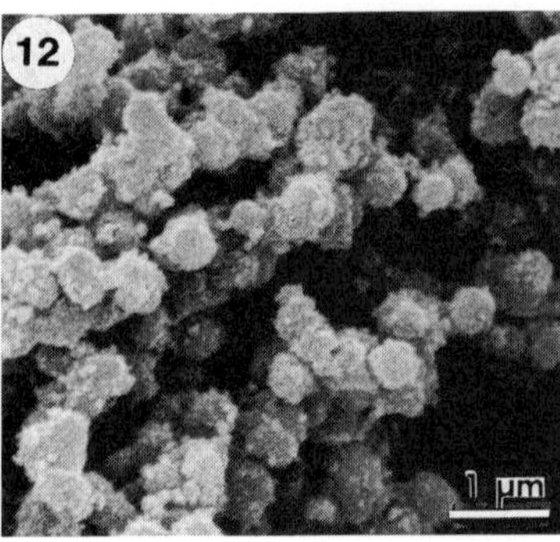

Figure 2.12 SEM micrograph of whey protein gels heated to 95°C (from Hermansson, 1986).

structure was shown to result in a significant increase in moisture loss (Hermansson, 1986). pH strongly influenced the water-holding properties of aggregated gels. Figures 2.14 and 2.15 show the difference in microstructure between whey protein gels formed at pH 6 and 8 respectively. As can be seen the gels of this whey protein concentrate were aggregated also at higher pH. As will be discussed, pure native β-lactoglobulin tends to form fine-stranded gels

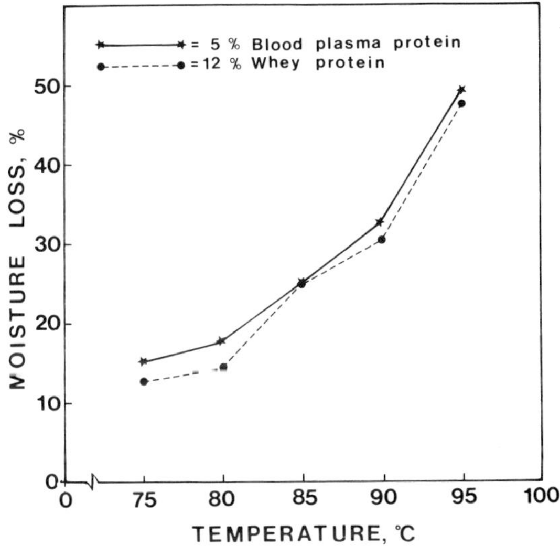

Figure 2.13 Moisture loss as a function of temperature of 5% blood plasma and 12% whey protein gels at pH 7 (from Hermansson, 1986).

Figure 2.14 SEM micrographs of whey protein gels formed at pH 6 (from Hermansson, 1986).

with increasing pH above the isoelectric region. However, whey protein systems are not pure and whey proteins are generally partially denatured prior to the onset of gelation. Figure 2.16 shows the moisture loss of whey and blood plasma protein gels as a function of pH. These results are consistent with observations that the coarser the network the poorer the water-holding capacity. It would be of interest to determine the pore-size distribution in these gels to establish a more

Figure 2.15 SEM micrographs of whey protein gels formed at pH 8 (from Hermansson, 1986).

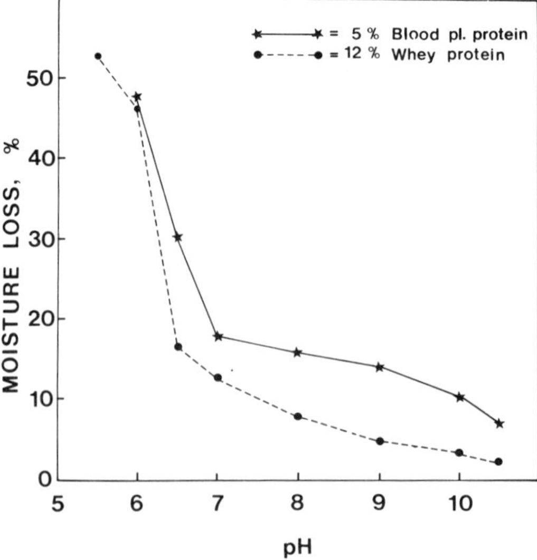

Figure 2.16 Moisture loss as a function of pH of 5% blood plasma protein and 12% whey protein heated solutions to 85°C (from Hermansson, 1986).

exact relationship between pore size and water-holding capacity. The possible use of nuclear magnetic resonance (NMR) imaging to quantify pore sizes > 10 μm has been discussed by Lillford *et al.* (1980). Indeed, image analysis of the finer pore size distribution of network structures may prove to be a powerful tool for the future.

2.4 Gel microstructure and rheological properties

There seems to be a direct relationship between the pore-size distribution and the water-holding capacity of protein gels. The relationship between gel micro-structure and rheological properties is not straightforward and small-deformation tests may provide different information than large-deformation tests. Studies have been carried out on β-lactoglobulin gels as a function of pH in which the microstructure as well as viscoelastic and fracture properties have been evaluated (Stading and Hermansson, 1990, 1991; Langton and Hermansson, 1992). Similar to other globular protein gels, β-lactoglobulin can form fine-stranded as well as aggregated particulate gels depending on pH and salt concentration (Clark and Lee-Tuffnell, 1986; Mulvihill *et al.*, 1990; Langton and Hermansson, 1992). In the absence of added salts, particulate gels were formed on heating at pH 4–6 and transparent fine-stranded gels formed below and above this pH range. The microstructure of the particulate gels could be separated into regular and irregular networks. The regular network gels formed at pH 5–6 were composed of particles with a narrow size distribution. An example of such a gel is shown by the SEM micrograph in Figure 2.17. At lower pH, more irregular net-works were formed in which the particles associated into larger spherical clusters, which were analogous to the appearance of a cauliflower. Figure 2.18 shows the uneven, rough microstructure of a fracture plane of such a gel visualized by SEM.

The shift from an aggregated to fine-stranded gel type is associated with a drastic change in the dimensions of the gel structure. Figure 2.19 shows the microstructure of a particulate gel imaged by light microscopy. The dimension of the 'threads' is in micrometres. In contrast, Figure 2.20 shows a micrograph of a fine-stranded network structure observed by TEM in which the 'strands' have dimensions in nanometres. Thus, there are enormous differences between partic-ulate and fine-stranded gels, but there are also differences between fine-stranded gels on either side of the pH range 4–6 (Langton and Hermansson, 1992).

Viscoelastic measurements are well-suited to monitor the gelation process. Stading and Hermansson (1990) found that it was possible to differentiate between fine-stranded and particulate β-lactoglobulin gels by viscoelastic measurements, but they could not be differentiated within the two gel types. The particulate gels were more strain-sensitive during gelation and more frequency dependent than the fine-stranded gels. They also had a higher storage modulus (G') than the fine-stranded gels at a constant β-lactoglobulin concentration of 12% (w/w).

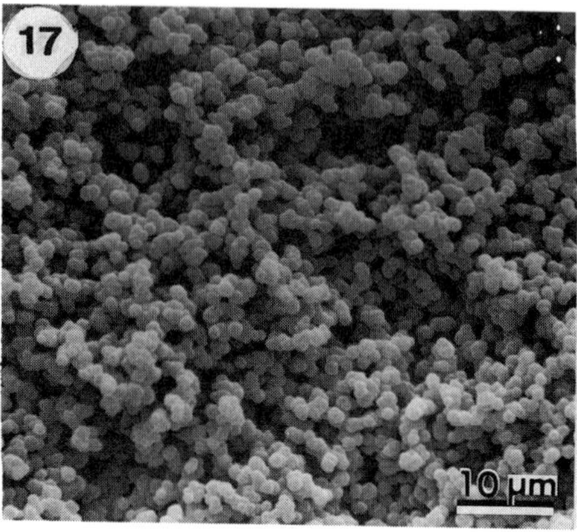

Figure 2.17 A regular particulate network of β-lactoglobulin.

Figure 2.18 An irregular particulate network of β-lactoglobulin.

Large deformation measurements permitted differentiation between regular and irregular particulate networks and between fine-stranded gels below and above the pH range for particulate gels (Stading and Hermansson, 1991). Stress–strain curves for 12% (w/w) β-lactoglobulin gels are shown in Figure 2.21. The particulate gels formed at pH 4.5, 5.5 and 6.0 exhibited similar strain at fracture but the stress at fracture differed. The regular gel structures formed at pH 5.5 and 6.0 had higher stress at fracture than the more uneven-type structure formed at pH 4.5. Large-deformation tests revealed large differences between fine-stranded gels formed at low and high pH. The gel formed at pH 3.0 was brittle and fractured at low strain and stress, whereas the gel formed at pH 7.5

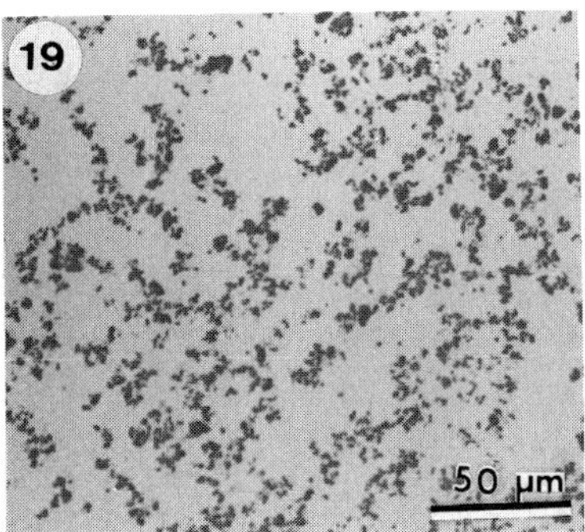

Figure 2.19 LM micrograph showing the gel structure of β-lactoglobulin at pH 5 (from Langton and Hermansson, 1991).

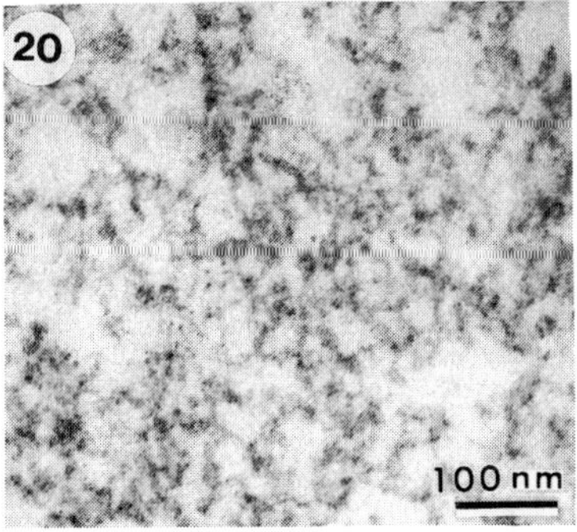

Figure 2.20 TEM micrograph showing the gel structure of β-lactoglobulin formed at pH 7 (from Langton and Hermansson, 1991).

had a high strain and a relatively high stress at fracture. Analysis of gel microstructure revealed that the fine strands were more curled with a longer contour-length between junctions at high pH than at low pH (Langton and Hermansson, 1992). This may contribute to the rubber-like behaviour at high pH, whereas the stiff strands at low pH could be responsible for the brittleness of these gels.

Figure 2.22 shows the change in Young's modulus (E) and the storage modulus (G') of a 12% (w/w) β-lactoglobulin gel as a function of pH. The scales were made so that E corresponded to 3G'. The most pronounced differences between E and G' were the two maximum values of E. It is interesting to note that these pH maxima coincide with the transitions between a particulate and fine-stranded gel structure. Microstructural analysis showed that the gel at the lower pH shift was composed of a mixture of particulate and fine-stranded structures. This was not the case for the sharper pH shift at pH 6.0, where either

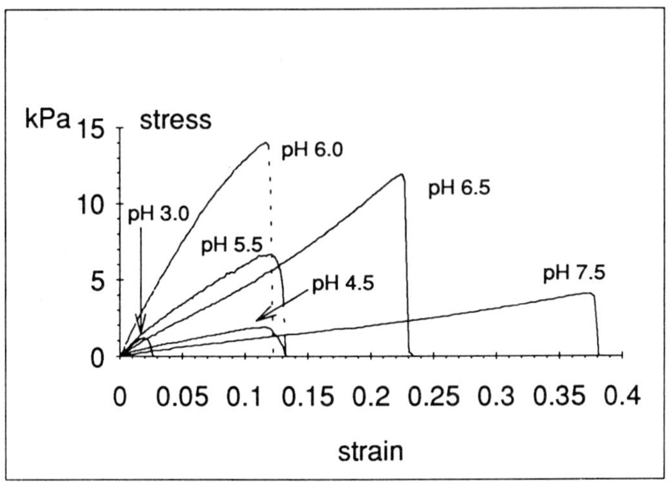

Figure 2.21 Stress–strain curves in tension for 12% β-lactoglobulin gels (from Stading and Hermansson, 1991).

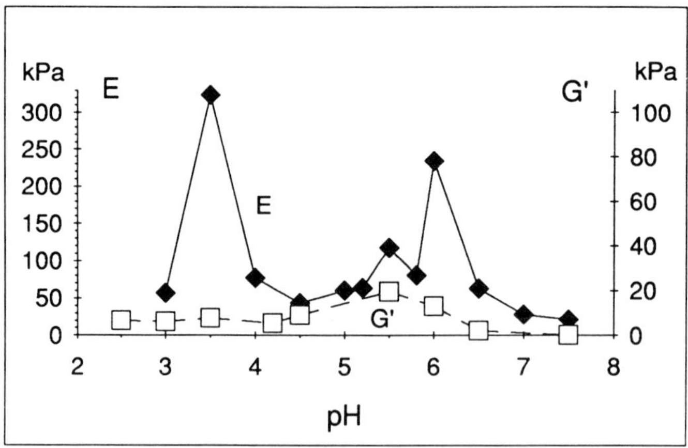

Figure 2.22 Young's modulus (E) and the storage modulus (G') of 12% β-lactoglobulin gels as a function of pH (from Stading and Hermansson, 1991).

a particulate or a fine-stranded structure was formed. On the particulate side of the pH shift, the stranded structure showed pronounced density fluctuations, which decreased with increasing pH. The microstructure of whey protein gels as a function of pH was similar to that of β-lactoglobulin, although the pH region for particulate gel formulation was wider.

The relationships between microstructure and rheological properties are complex and more work is needed in this area. At present, it seems as if tensile tests reveal more details about gel microstructure than viscoelastic measurements, whereas the latter are well-suited for increasing our understanding of gelation mechanisms.

2.5 Image analysis—a potential for the future

The ability to quantify structural information by image analysis strengthens electron microscopy as a tool for the evaluation of structure–function relationships of food biopolymers. Structural parameters that are of importance for functional properties can be estimated using this technique. This opens up new possibilities for mathematical modelling from real structures rather than from model systems, which rely on simplified assumptions. Calculations can be based on a large number of micrographs and further processed for statistical evaluation.

A software package developed in collaboration with Context Vision (Linköping, Sweden) for food biopolymer and gel structures includes the following parameters: (i) flexibility of strands; (ii) thickness of strands; (iii) length of strands between junctions; (iv) frequency of junction points; (v) number of strands in a junction; (vi) angles between strands in a junction; and (vii) pore size. The flexibility of single molecules and supramolecular structures can be measured, as well as the flexibility of strands between two junction points in a network. The flexibility is determined by calculating the ratio of the total length of a strand (l_o) to the shortest distance between the two ends (l_a). The thickness of strands is measured perpendicular to the longitudinal axes i.e. elongation. This is not as trivial for the computer as it is for the human eye. The frequency of junction points is an interesting parameter since it gives a measure of the density of the network and can be used for the determination of density fluctuations of a network structure. The pore size distribution also provides a measure of the network density, although this parameter is more of an approximation. In the case of TEM micrographs, where the gel is represented in one plane, the pore size is approximated from the largest circle that can be enclosed in a void within the boundaries of the strands.

The information that can be obtained by image analysis of gel structures can be illustrated using data from soy glycinin gels. Soy glycinin is composed of acidic and basic subunits, which alternate in a hexagonal structure. In the native molecules, 12 subunits form a two-layered hexagonal structure (Badley *et al.*, 1975). The isoelectric region is the pH range 4.75–5.40 for the acidic subunits and in the pH range 8.0–8.5 for the basic subunits. For the purposes of this work, gels were studied in the pH range 4–9. No gels formed at pH 5, i.e. in the isoelectric region of the acidic subunits. However, at pH 6 an aggregated gel was obtained with large compact aggregates. The fine-stranded gels formed at pH 7 were studied in detail and the gel structure of soy glycinin in distilled water is shown in Figure 2.23. The strands of this gel have a hollow cylindrical structure

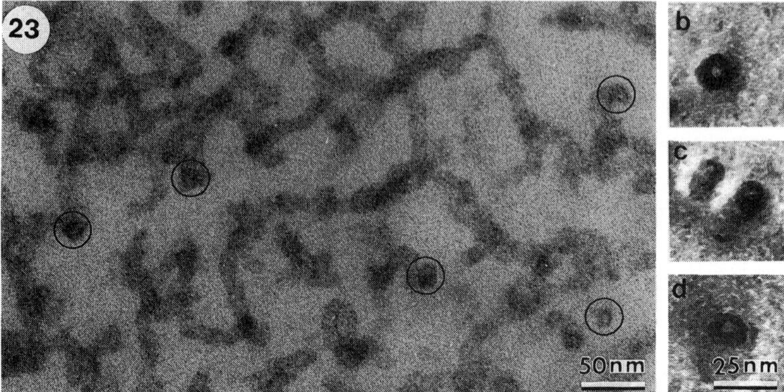

Figure 2.23 (a) TEM micrograph of a thin section of a glycinin gel made in distilled water at 95°C. Some cross-sections of strands are encircled; (b–d) Cross-sections of the hollow cylindrical glycinin strands at high magnification prepared by freeze-etching and rotary shadowing (from Hermansson, 1985).

with a spiral arrangement of subunits of 10–15 nm diameter, similar to that of the native quarternary structure (Hermansson, 1985). The strands of gels formed at higher pH had thinner strands. Table 2.2 shows the flexibility and the distance between junction points of soy glycinin gel networks formed at pH 4, 7, 8 and 9. The flexibility ratio (Table 2.2) indicates that the gel formed at pH 7 had the most flexible strands. The distance between the junction points was twice as long for the gel made at pH 7 than for gels made at pH 4, 8 and 9; this gel therefore has a more open structure. The same information was obtained from measurements of the frequency of junction points and of the pore-size distribution (data not shown).

Results from image analysis demonstrated that the soy glycinin gels formed at pH 7 had a completely different structure from the gels formed at pH 4, 8 or 9. The results are interesting since pH 7 is between the isoelectric points of the acidic and basic subunits. The gels made at other pH values thus have a different mode of aggregation into strands and a different type of gel structure than at pH 7. The results illustrate how image analysis can be used to advantage in evaluating network structures. There is potential for the use of

Table 2.2 Some structural parameters of soy glycinin gels at pH 4–9 derived by image analysis (Langton and Hermansson, submitted)

	pH			
Structural parameters	4	7	8	9
Flexibility ratio l_a/l_0	0.79	0.72	0.79	0.80
Distance between junction points (nm)	26	57	27	25

image analysis in the measurement of more complex structural parameters such as density fluctuations, distribution of network structures in mixed gels, as well as various phases in multiphase systems. At present these parameters are difficult to quantify using other techniques. By combining measuring techniques it should be possible to make satisfactory predictions of the relationships between the microstructure of food biopolymers and their functional properties.

References

Badley, R.A., Atkinson, O., Hanser, H. (1975) The structure, physical and chemical properties of the soy bean protein glycinin. *Biochem. Biophys. Acta.* **412**:214–28.

Clark, A.H. and Lee-Tuffnell, D.D. (1986) Gelation of globular proteins, in: *Functional Properties of Food Macromolecules*, (eds J.R. Mitchell and D.A. Ledward), Elsevier Applied Science, London, pp.203–72.

Clark, A.H. and Ross-Murphy, S. (1987) Structural and mechanical properties of biopolymer gels. *Adv. Polymer Sci.* **83**:57–192.

Elliot, A. and Offer, G. (1978) Shape and flexibility of myosin molecules. *J. Mol. Biol.* **123**:505–19.

Hermansson, A.-M (1985) Structure of soya glycinin and conglycinin gels. *J. Sci. Food Agric.* **36**:822–32.

Hermansson, A.-M. (1986) Water and fat-holding, in: *Functional Properties of Food Macromolecules*. (eds J.R. Mitchell and D.A. Ledward), Elsevier Applied Science, London, pp. 273–314.

Hermansson, A.-M (1988) Gel structure of food biopolymers, in: *Food Structure: Its Creation and Evaluation*. (eds J.M.V. Blanshard and J.R. Mitchell), Butterworths, Borough Green, Sevenoaks, pp. 25–40.

Hermansson, A.-M. and Langton, M. (1988) Filamentous structures of bovine myosin in diluted suspensions and gels. *J. Sci. Food Agric.* **42**:335–69.

Hermansson, A.-M. and Langton, M. (1992) Electron microscopy, in: *Physical Techniques for the Study of Food Biopolymers*, (ed. S.B. Ross-Murphy), Blackie Academic & Professional, Glasgow, pp. 277–342.

Hermansson, A.-M., Langton, M. and Harbitz, O. (1986) Formation of two types of gels from myosin. *J. Sci. Food Agric.* **37**:69–84.

Huxley, H.E. (1963) Electron microscope studies on the structure of natural and synthetic protein filaments from striated muscle. *J. Mol. Biol.* **7**:281–308.

Kaminer, B. and Bell, A.L. (1966) Myosin filamentogenesis: effects of pH and ionic concentration. *J. Mol. Biol.* **20**:391–401.

Langton, M. and Hermansson, A.-M. (1992) Fine-stranded and particulate gels of β-lactoglobulin and whey protein at varying pH. *Food Hydrocoll.* **5**:523–39.

Lillford, P.J., Clark, A.H. and Jones, D.V. (1980) Distribution of water in heterogeneous food and model systems, in: *Water in Polymers*. (ed. S.P. Rowland), *ACS Symposium Series No, 127*, pp. 178–95.

Luyten, H. (1988) *The Rheological and Fracture Behaviour of Gouda Cheese*. PhD Thesis, Wageningen Agricultural University, Wageningen, The Netherlands.

Mulvihill, D.M., Rector, D. and Kinsella, J.E. (1990) Effects on structuring and destructuring anions on the rheological properties of thermally induced β-lactoglobulin gels. *Food Hydrocoll.* **4**:267–76.

Ross-Murphy, S.B. (1984) Rheological methods. *Crit. Rep. Appl. Chem.* **5**:103–37.

Stading, M. and Hermansson, A.-M. (1990) Viscoelastic behaviour of β-lactoglobulin gel structures. *Food Hydrocoll.* **4**:121–35.

Stading, M. and Hermansson, A.-M. (1991) Large deformation properties of β-lactoglobulin gel structures. *Food Hydrocoll.* **5**:339–52.

Walker, M., Knight, P. and Trinick, J. (1985) Negative staining of myosin molecules. *J. Mol. Biol.* **184**:535–42.

Yamamoto, K.Y. (1990) Electron microscopy of thermal aggregation of myosin. *J. Biochem.* **108**:896–98.
Yamamoto, K.Y., Samejima, K. and Yasui, T. (1988) Heat-induced gelation of myosin filaments. *Agric. Biol. Chem.* **52**:1801–11.
Ziegler, R. and Foegeding, E.A. (1990) Gelation of proteins. *Adv. Food Nutr. Res.* **34**:203–98.

3 Probing structural changes and preparation of protein domains by limited proteolysis

H.E. SWAISGOOD, S.X. CHEN, S. OH AND G.L. CATIGNANI

Abstract

Many food proteins, in addition to the muscle proteins, are likely to have multiple domains, e.g. milk proteins and egg proteins have been shown to yield large polypeptide fragments upon limited proteolysis. Probing for domain structure and preparation of protein domains by limited proteolysis is especially advantageous with immobilized proteinase bioreactors, since autolysis is prevented and the degree of hydrolysis is easily and precisely controlled. Thus, preliminary studies with immobilized trypsin suggest that the central β-barrel domain of β-lactoglobulin can be obtained by limited proteolysis. By analogy to the evolution of new biological functions by recombination of structural domains or motifs, it is suggested that new food functionality properties may also be elicited by recombination. Moreover, such recombination, although random, could possibly be achieved by limited cross-linking as catalyzed by transglutaminase. An active, immobilized form of this enzyme has been prepared and the kinetics of its action with β-lactoglobulin and α_{s1}-casein determined.

3.1 Introduction

The tertiary structure of many proteins is composed of assemblies of structural domains containing motifs that have been recruited in evolution for specific biological functions. Owing to numerous intradomain interactions, such domains are not very susceptible to proteolysis; however, the interdomain peptide segments exhibit much greater flexibility and are thus very susceptible to enzymatic cleavage. In recent years, numerous examples of domain separation by limited proteolysis have been reported and, in fact, limited proteolysis is now used as a tool to probe the domain structure of multidomain proteins such as phosphoglycerate kinase, calcineurin A, myosin, tropomyosin, streptococcal M protein and von Willebrand factor.

3.2 Protein structure and its relationship to proteolysis

The rate of enzyme-catalyzed hydrolysis of a particular peptide bond is determined by two major factors, i.e. the susceptibility of that bond to the specific proteinase in question and the flexibility of the protein chain in the region of the bond. In other words, the phi and psi angles of the susceptible bond and surrounding residues must fluctuate to allow a high-affinity fit of the substrate chain into the surface topology of the active site of the enzyme, thus bringing the catalytic residues into proper orbital alignment for covalent bond rearrangement to occur. Linderstrom-Lang (1952) first recognized the importance of substrate protein structure to proteolysis rates and this phenomenon has since been characterized and adapted to probe substrate proteins for their structure (Rupley, 1967; Matthyssens *et al.*, 1972; Mihalyi, 1978; Burgess *et al.*, 1975; Church *et al.*, 1982a,b; Ueno and Harrington, 1984; Swaisgood and Catignani, 1987; Swaisgood, 1989). Quantitatively, in the pseudo first-order kinetic region, the relationship between velocity of peptide bond hydrolysis (v) and structure can be written (Rupley, 1967; Matthyssens *et al.*, 1972; Church *et al.*, 1982b; Swaisgood and Catignani, 1987):

$$v = \frac{k_c \, [E]^\circ}{K_m^{app}} f_D \, [P]^\circ = k f_D \, [P]^\circ \qquad (3.1)$$

where k is a pseudo first-order rate constant; f_D is the fraction of substrate protein molecules with unfolded conformation in the susceptible region; $[P]^\circ$ is the total substrate protein concentration; k_c is the catalytic rate constant; K_m^{app} is the Michaelis constant; and $[E]^\circ$ is the total enzyme concentration. If the assumption is made that an equilibrium exists between unfolded and native conformations in a two-state transition, the equation can be written:

$$v = k \, (1 - f_D) \, K_{conf} \, [P^\circ] \qquad (3.2)$$

where K_{conf} represents the conformational equilibrium constant.

Considerable experimental evidence to support the above principles has been obtained and, in many cases, even the assumption of a two-state transition equilibrium. A conclusive illustration of the importance of structure to proteolytic susceptibility comes from the demonstration that limited proteolysis or autolysis of thermolysin occurs specifically in those regions of the protein chain that are characterized by the highest mobility (flexibility), as given by the crystallographic temperature factors (Fontana *et al.*, 1986). There are numerous examples of the use of proteinases to probe conformational transitions in protein structure. The thermal transition of lysozyme at pH 2 was analyzed from rates of pepsin-catalyzed hydrolysis (Matthyssens *et al.*, 1972). Coincidence of these data with those derived from optical rotation and lysozyme activity indicated that the thermal unfolding is represented by a two-state transition. Other examples include the use of trypsin to probe conformational changes in the flexible region of the α-helical-coiled coil in the myosin rod (Ueno and Harrington, 1984) and the flexible region of the coiled-coil structure of tropomyosin (Ueno, 1984). Also, ligand-modulated conformational changes in proteins have been probed with proteinases, such as the binding of substrate aspartate and effectors ATP and CTP, to aspartate transcarbamylase monitored with trypsin, subtilisin or Pronase

(McClintock and Markus, 1968); conformational changes have been induced in adenosinetriphosphatase by ADP, ATP or phosphate probed with trypsin (Di Pietro *et al.*, 1983) and structural changes have been achieved in carbamyl phosphate synthetase I, resulting from binding of the activator N-acetylglutamate and monitored by limited proteolysis with elastase (Marshall and Fahien, 1988).

3.3 Use of immobilized proteinases as structural probes

Several advantages are realized by using immobilized forms of proteinases for probing protein structures for flexibility, as follows:

- Extent of reaction is easily and precisely controlled
- An enzyme denaturation step is not required
- Hydrolysis products do not require purification to remove enzyme or autolysis products
- Autolysis does not occur
- Stability of the enzyme may be greater than that of soluble forms.

The primary advantages are associated with the precise control over the extent of reaction without a denaturation step and the prevention of autolysis. However, for characterization of structural transitions resulting from thermal treatment or exposure to denaturants, the potential for increasing the stability of the proteinase (perhaps also related to autolysis) is also quite important. For example, carboxypeptidase A immobilized on Sephadex was used to follow the thermal unfolding of the C-terminal domain of ribonuclease A, yielding results similar to that obtained by spectral methods (Burgess *et al.*, 1975). A slight deviation of the spectral data from the proteinase data was observed in the low-temperature range of the transitions reflecting the slightly lower stability of the N-terminal domain containing Tyr_{25} and Tyr_{92}. Such studies were not possible with soluble carboxypeptidase A since it is unstable above 60°C, whereas the immobilized enzyme was stable between 60 and 70°C.

To probe for general changes in protein structure it is desirable to use a proteinase with broad substrate specificity or a mixture of proteinases to ensure that a structural change in a particular region will not be missed due to the lack of a susceptible sequence. We have used immobilized Pronase, a mixture of proteinases from *Streptomyces griseus*, to monitor urea-induced structural transitions in several milk proteins (Church *et al.*, 1982b; Swaisgood and Catignani, 1987; Swaisgood, 1989). At pH 7.5 the midpoints of the urea-induced unfolding transitions as measured by proteolysis with immobilized Pronase are roughly 1 M for whole caseins, 4 M for β-lactoglobulin and 7.2 M for lysozyme. Thus, immobilized Pronase retained sufficient activity in 8 M urea to permit accurate measurement of the rate of hydrolysis (Burgess *et al.*, 1975; Church *et al.*, 1982b); furthermore, the decrease in enzyme activity at this concentration was reversible (Church *et al.*, 1982a).

A criticism frequently directed at the use of immobilized enzymes is that they continually leak from the support. However, this problem is not observed if the enzyme is immobilized via a stable covalent bond and if the support is thoroughly washed to remove all noncovalently adsorbed enzyme. Stable bioreactors have been prepared by immobilization to succinamidopropyl-silica through formation of an amide linkage (Swaisgood *et al.*, 1976; Church *et al.*, 1982a; Janolino and Swaisgood, 1982; Porter *et al.*, 1984 Swaisgood and Horton, 1989). By comparison, the commonly used method of immobilization on cyanogen bromide-activated Sephadex does not result in a stable covalent bond (Kohn and Wilchek, 1982). Furthermore, enzymes immobilized by a stable covalent bond can be washed extensively with 4 M urea, which removes all adsorbed protein and in most cases will not cause any loss of activity. Bioreactors of immobilized proteinases so constructed have been used in laboratories for long periods without detectable loss of activity.

3.4 Probing protein domain structure with proteinases

Most proteins are assemblies of substructures of domains, these regions being defined as a group of residues with a minimum surface–volume ratio resulting from many interactions between residues within the domain but few interactions with residues outside the domain (Neurath, 1986). The interdomain regions thus represent 'hinges' or flexible residues in the protein backbone. Consequently, the interdomain regions are most susceptible to proteolysis. Indeed, limited proteolysis is an effective method of isolating and identifying protein domains (Neurath, 1986). The recognition of protein domains, together with the observation that homologous domains are found in structurally or functionally unrelated proteins, has added a new dimension to the understanding of protein evolution.

2.4.1 Proteolytic analysis of domains in globular proteins

Immobilized *Staphylococcus aureus* V8 protease has been used to probe the domain structure of phosphoglycerate kinase (PGK) (Betton *et al.*, 1989). In physiological buffers PGK is not very susceptible to proteolysis; therefore, low concentrations of guanidine hydrochloride (Gdn · HCl) were used to increase the flexibility of the interdomain region. A 4 h digestion at 12°C yielded 25 kD and 11 kD fragments at appropriate Gdn · HCl concentrations. Densitometric analysis of SDS-polyacrylamide gel electrophoresis (PAGE) indicated that the 11 kD fragment was very stable at all Gdn · HCl concentrations, but that the 25 kD domain was produced in optimal amounts in 0.7 M Gdn · HCl (Figure 3.1). Progress curves of the digestion (Figure 3.2) indicated that the 25 kD domain was stable under the latter conditions. Further analysis by CD spectroscopy and amino-acid sequencing showed that cleavage occurred at Glu_{192} Leu_{193} yielding the C-terminal domain, Leu_{193} Val_{416} (Figure 3.3), and that this 25 kD domain

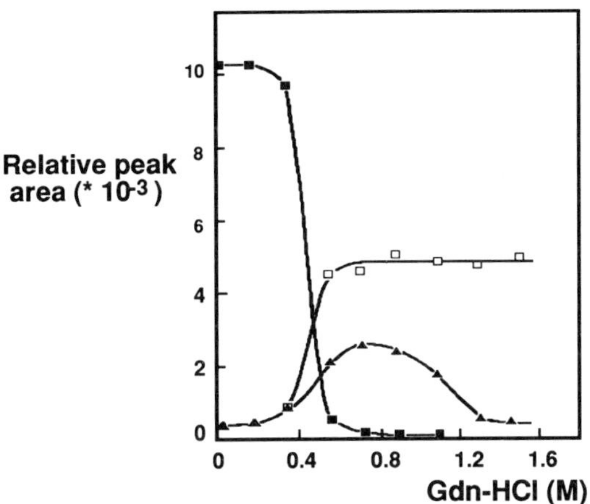

Figure 3.1 Dependence of the relative quantities of proteolysis products upon guanidine hydrochloride concentration. Relative areas of intact phosphoglycerate kinase (■), 25 kD fragment (▲) and 11 kD fragment (□) were obtained from densitometric analysis of SDS-polyacrylamide gel electrophoresis. Limited proteolysis was performed with immobilized *S. aureus* V8 protease at pH 7.5 for 4 h at 12°C (from Betton *et al.*, 1989, with permission).

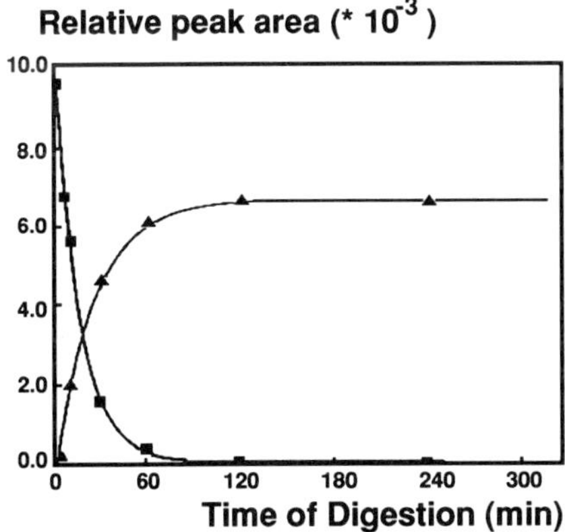

Figure 3.2 Progress curves for limited proteolysis of phosphoglycerate kinase (PGK) with immobilized *S. aureus* V8 protease in 0.7 M guanidine hydrochloride. Relative areas of native PGK (■) and 25 kD fragment (▲) were obtained from densitometric analysis of SDS-gel electrophoresis. Conditions were as in Figure 3.1 (from Betton *et al.*, 1989, with permission).

had a similar structure to that of native PGK with the exception of a slightly lower helix content (Betton *et al.*, 1989). Thus, proteolysis occurs in the less-stable helix α7 connecting the N-terminal and C-terminal domains.

Calcineurin, the major calmodulin-binding protein phosphatase in the brain, is

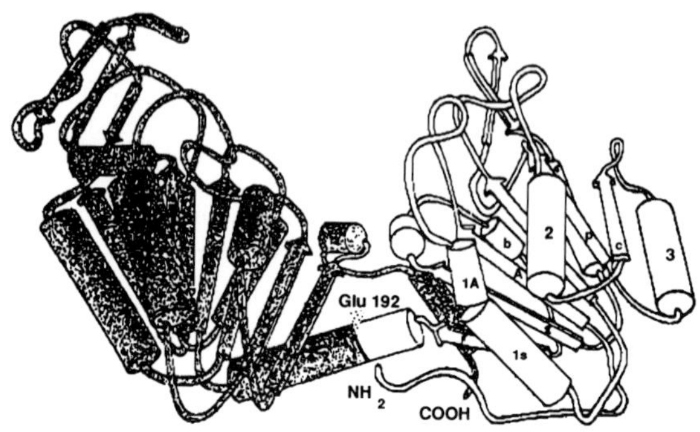

Figure 3.3 Three-dimensional structure of phosphoglycerate kinase depicting the site of limited proteolysis with immobilized *S. aureus* V8 protease (from Betton *et al.*, 1989, with permission).

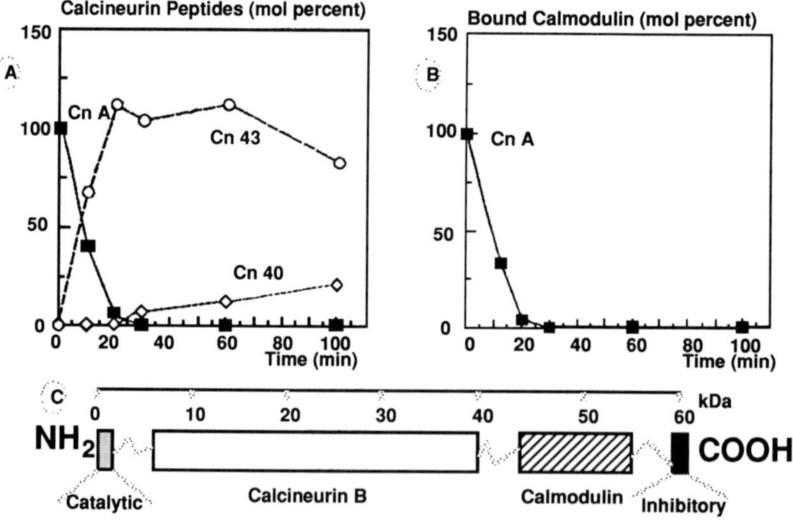

Figure 3.4 Limited proteolysis of calcineurin by clostripain in the absence of calmodulin. (A) Progress curves for proteolysis as determined by densitometric analysis of SDS-polyacrylamide gel electrophoresis. Relative amounts of calcineurin A (■), 43 kD fragment (○) and 40 kD fragment (◇). (B) Binding of ^{125}I-calmodulin to components in the SDS-gel. (C) Domain structure of calcineurin A deduced by proteolytic probing indicating the catalytic, calcineurin B binding, calmodulin binding and inhibitory domains (Adapted from Hubbard and Klee, 1989, with permission.)

a heterodimer of calcineurin A (61 kD) and calcineurin B (19 kD). The domain structure of the larger subunit, calcineurin A has been elucidated with clostripain, an Arg-specific protease, as a proteolytic probe (Hubbard and Klee, 1989). In the absence of calmodulin, calcineurin A disappears rapidly with the concomitant appearance of a 43 kD domain, which is more slowly hydrolyzed to a 40 kD domain (Figure 3.4a). Loss of calmodulin binding ability coincides with the disappearance of calcineurin A (Figure 3.4b). The 43 kD and 40 kD domains retain the ability to bind calcineurin B and exhibit enhanced enzymatic activity. By contrast, in the presence of calmodulin, proteolysis first yields a 57 kD polypeptide with a concomitant ten-fold increase in activity (Figure 3.5a). This polypeptide is more slowly hydrolyzed to give a 55 kD polypeptide that still binds calmodulin (Figure 3.5b) and calcineurin B, but most of the enzyme activity is lost. Longer periods of proteolysis (>5 h) lead to the conversion of the 55 kD polypeptide to 14 kD and 42 kD domains. The 14 kD domain binds calmodulin, whereas the 42 kD domain binds calcineurin B. Such analyses of the products of limited proteolysis indicate that calcineurin A is composed of four domains as illustrated in Figures 3.4c and 3.5c. This study also illustrates the use of ligand binding to enhance the stability and thus protect a specific domain.

3.4.2 Proteolytic analysis of domains in coiled-coil rod proteins

The ability of proteinases to release domains of myosin has been recognized for many years (McLachlin, 1984; Rawn, 1989). The α-helical coiled-coil rod of

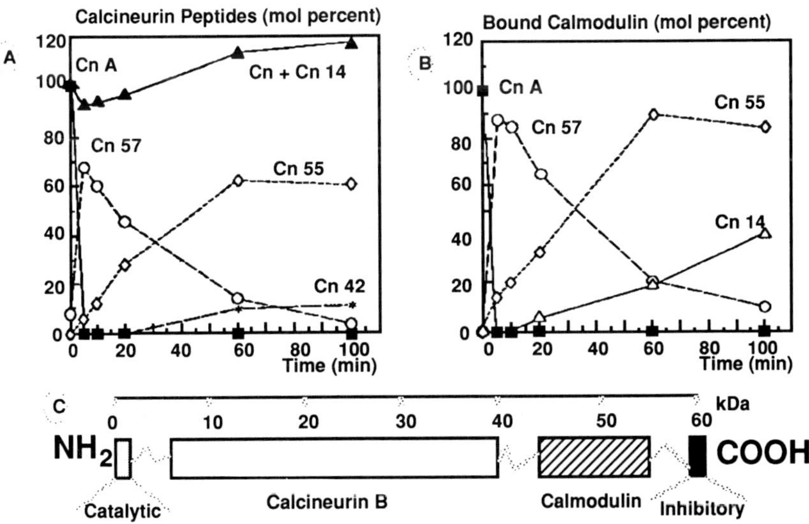

Figure 3.5 Limited proteolysis of calcineurin by clostripain in the presence of calmodulin. (A) Progress curves for proteolysis as determined by densitometric analyses of SDS-polyacrylamide gel electrophoresis. Relative amounts of calcineurin A (■) and its proteolytic fragments 57 kD (O), 55 kD (◇), 42 kD (*) and the sum of calcineurin B and 14 kD (▲). (B) Binding of ^{125}I-calmodulin to components in the SDS-gel. (C) Domain structure of calcineurin A as deduced from limited proteolysis indicating the catalytic, calcineurin B binding, calmodulin binding and inhibitory domains. (Adapted from Hubbard and Klee, 1989, with permission.)

myosin contains a hinge region with lower stability, i.e. more flexibility, and is thus susceptible to proteolysis (Rawn, 1989). Stability in this region has been probed with trypsin, chymotrypsin and papain (Ueno and Harrington, 1984) revealing that at low temperatures (5–25°C) cleavage is restricted to a narrow region separating the light meromyosin (LMM) and heavy meromyosin (HMM) domains. At higher temperatures (25–40°C) a much longer segment of structure 'melts' including part of the S-2 domain, which is the rod segment between the LMM/HMM junction and the globular head (S-1 domain). The 'swivel' region joining the globular S-1 domain to the S-2 segment of HMM is very susceptible to cleavage with papain (McLachlin, 1984; Rawn, 1989).

Similarly, trypsin has been used to probe a flexible region of the α-helical coiled-coil tropomyosin structure (Ueno, 1984). The 33 kD rod is rapidly hydrolyzed at a flexible segment near the middle (Figure 3.6). Following cleavage, the N-terminal domain is apparently unstable since it was not found. The specific site(s) of hydrolysis exhibits a temperature dependence that is reflective of slightly different structural stabilities in this flexible region. Most interestingly, formation of an interchain disulfide bond at Cys_{190} greatly increases the susceptibility of a nearby peptide bond, resulting in release of the C-terminal 12 kD domain C_3 (Figure 3.7).

A cell surface, major antiphagocytic determinant produced by Group A *Streptococci* (streptococcal M protein) is another interesting α-helical coiled-coil protein. This protein consists of a short nonhelical N-terminal domain connected by an extended coiled-coil domain to a C-terminal cell anchor domain (Khandke *et al.*, 1990). Probing of the structure of the coiled-coil rod revealed a flexible junction in this domain that is susceptible to pepsin at pH 5.5, which corresponds

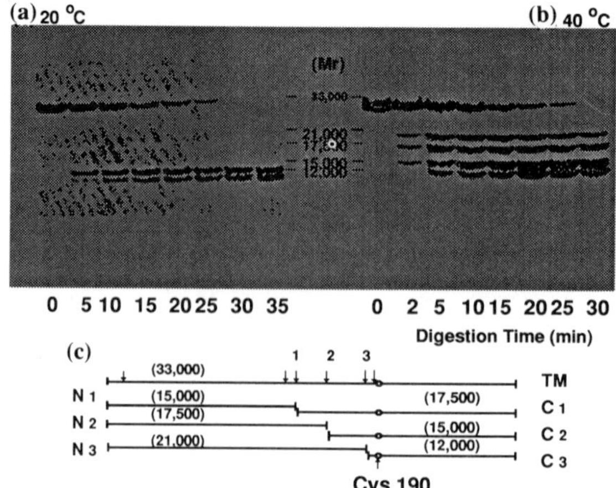

Figure 3.6 Limited proteolysis of tropomyosin by trypsin analyzed by SDS-polyacrylamide gel electrophoresis. Results of digestion of reduced tropomyosin at pH 7.05 in 20 mM cacodylic acid, 1 mM EDTA, 0.25 M NaCl and 1 mM β-mercaptoethanol at (a) 20°C and (b) 40°C. Sites of cleavage in the coiled-coil rod are noted (c). (Adapted from Ueno, 1984, with permission.)

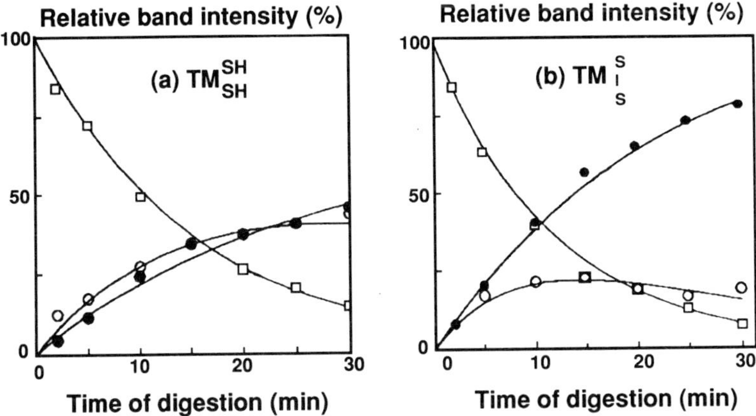

Figure 3.7 Effect of disulfide bond formation on the limited proteolysis of tropomyosin by trypsin. Progress curves obtained by densitometric analyses of SDS-polyacrylamide gel electrophoresis. Relative amounts are given for native tropomyosin (□), sum of C-terminal 17.5 and 15 kD fragments (○), and the 12 kD fragment C$_3$ (●). (Adapted from Ueno, 1984, with permission.)

to the boundary between subdomains II and III (Khandke *et al.*, 1990). Specific cleavage in this region permitting isolation of the domain containing subdomains I and II is very sensitive to the pH of limited proteolysis. Other structural studies indicated that the coiled-coil domain is composed of three subdomains named I, II and III (Fischetti *et al.*, 1988). The hinge region between variable (I and II) and conserved (III) subdomains of streptococcal M proteins is observed in the M protein of all serotypes. It thus appears that a hinge region may be important to the function of many types of α-helical coiled-coil proteins both from prokaryotes and eukaryotes.

3.4.3 Proteolytic analysis of a multifunctional blood protein

Immobilized *Staphylococcus aureus* V-8 protease has been used to probe the domain structure of von Willebrand factor (vWF) (Girma *et al.*, 1986). This multidomain protein participates in the adhesion of platelets to the sub-endothelium and in platelet aggregation. Analysis of limited proteolysis of the reduced protein shows a rapid and specific cleavage (a Glu–Glu bond) of the 270 kD protein to produce 110 kDa domains as determined by SDS-PAGE (Girma *et al.*, 1986; Figure 3.8). A minor cleavage site occurs in the 170 kD domain, which gives a 50 kD fragment. Native vWF is a multimer in which the 270 kD polypeptide chains are linked head-to-head and tail-to-tail by disulfide bonds (Figure 3.9). Thus cleavage with immobilized V-8 protease yields dimers of the 170 kD domains (fragment III) and the 110 kD domains (fragment II). Fragment III represents the N-terminal domain of the original 270 kD vWF and binds to platelets while fragment II, representing the C-terminal domain, does not.

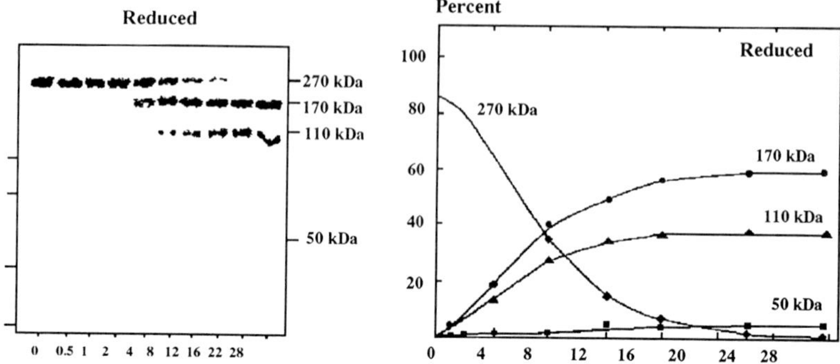

Figure 3.8 Limited proteolysis of human blood von Willebrand factor with immobilized *S. aureus* V8 protease. Digestion was performed in 0.05 M NaCl and 0.01 M EDTA at room temperature. SDS-polyacrylamide gel electrophoresis included β-mercaptoethanol and was analyzed by densitometry. (Adapted from Girma *et al.*, 1986, with permission.)

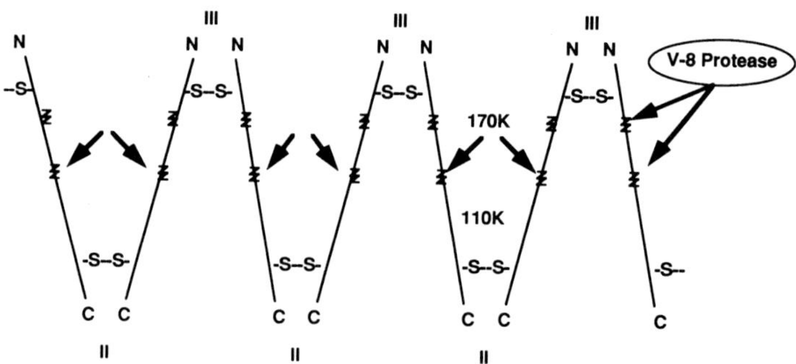

Figure 3.9 Proposed model for limited proteolysis of von Willebrand factor (vWF) by immobilized *S. aureus* V8 protease. Each vWF multimer is composed of a linear arrangement of 270 kD subunits linked head-to-head and tail-to-tail by disulfide bridges. Primary cleavage occurs between two glutamyl residues yielding a C-terminal homodimer of the 110 kD fragment (fragment II) and an N-terminal homodimer of the 170 kD fragment (fragment III). A minor secondary cleavage at a Glu–Gly sequence in the C-terminal region of the 110 kD fragment produces a 50 kD fragment (fragment I). (Adapted from Girma *et al.*, 1986, with permission.)

3.5 Possible domains in food proteins

The multidomain structure of proteins allows a single molecule to have multiple biological functions, thus permitting modulation of functions that otherwise would not be possible if each existed as a separate molecular entity. Undoubtedly, the same is true for protein structure and food functionality. A multidomain protein with different domain stabilities, capable of interacting with several other molecular entities in the food, offers a more diverse range of functionalities than would a molecule capable of only one interaction. Evidence suggests that many food proteins are assemblies of more than one domain, which thus adds another dimension to the potential for design of functionality. The well-established multidomain structures of several muscle proteins has already been discussed. In this section, the evidence for multidomain structure in several other food proteins will be examined.

Limited proteolysis of β-casein (CN) occurs naturally in milk due to the inherent presence of plasmin (Eigel, 1977). The sites of proteolysis are in the region separating the proposed hydrophobic and polar domains (Swaisgood, 1982) generating β-CN X(f29–209), β-CN X(f106–209) and β-CN X(f108–209), known as the γ-caseins, as well as β-CN (f1–105 or f1–107), β-CN (f29–105 or f29–107) and β-CN (f1–28). Another historical example of limited proteolysis, which is actually an ancient bioprocessing operation, is the specific cleavage of κ-casein between Phe_{105} Met_{106} by chymosin liberating a proposed polar domain (Swaisgood, 1982). This causes a dramatic change in the functionality of the remaining hydrophobic domain, κ-CN (f1–105) and results in curd formation.

Two peptides, β-CN (f1–28) and β-CN (f1–52), obtained by limited proteolysis with plasmin and α-chymotrypsin, respectively, have been isolated and structurally characterized (Chaplin *et al.*, 1988). These peptides are derived from the N-terminal polar domain of β-casein; hence, the relationship between their structure and functionality should be informative. Analysis of their secondary structures by CD spectroscopy indicated that both contained some β-structure in aqueous solutions and that β-CN (f1–52) exhibited a propensity to form α-helix, whereas β-CN (f1–28) did not form any α-helix even in helix-inductive organic solvents (Chaplin *et al.*, 1988). Furthermore, β–CN (f1–52), which may represent the polar domain of β-casein, is capable of forming an amphipathic α-helix encompassing residues 23–37 and predictions from the sequence indicate a high probability of α-helix including residues 25–46. Such an amphipathic helix may have important interfacial properties. On the other hand, β-CN (f1–28), which also represents the N-terminal portion of β-CN (f1–52), apparently has no helix-forming tendencies and has a high charge density (Swaisgood, 1982; Chaplin *et al.*, 1988).

Limited proteolysis of α_{s1}-casein with pepsin (Shimizu *et al.*, 1986) or chymosin (Kaminogawa *et al.*, 1980; Creamer *et al.*, 1982) yields α_{s1}-CN (f1–23) and α_{s1}-CN (f25–199). The N-terminal peptide, α_{s1}-CN (f1–23), represents a somewhat hydrophobic domain of the parent protein (Swaisgood, 1982), including part of a probable hydrophobic β-sheet structure (Kumosinsky *et al.*, 1991). This peptide is completely soluble in the pH range 3–9 and exhibited rather good emulsifying properties in the pH range 3–6, whereas α_{s1}-casein is rather insoluble (Shimizu *et al.*, 1986).

The C-terminal portion of α_{s1}-casein resulting from limited proteolysis with chymosin, α_{s1}-CN (f25–199), contains both a proposed polar domain and a rather

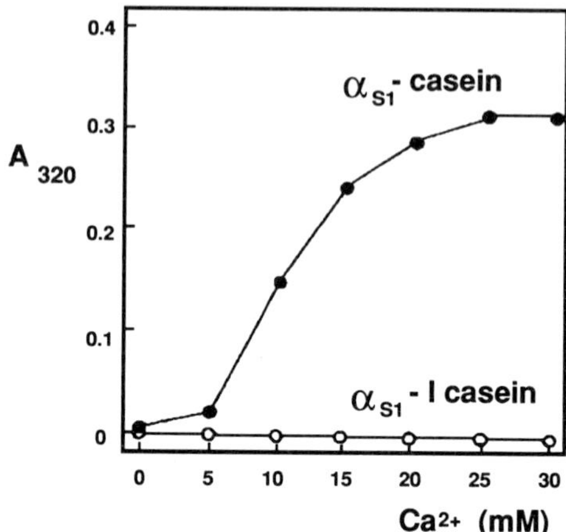

Figure 3.10 Sensitivity of α_{s1}-CN(f25–199) compared with the parent protein, α_{s1}-casein, to precipitation by calcium ion (from Kaminogawa *et al.*, 1980, with permission).

large hydrophobic domain (Swaisgood, 1982). This polypeptide (also known as α_{s1}-I) is missing the rather hydrophobic N-terminal domain and consequently exhibits reduced surface hydrophobicity (Creamer *et al.*, 1982). Therefore, the solubility of α_{s1}-I in the presence of calcium ions is greatly different from α_{s1}-casein even though it binds the ion to the same extent (Kaminogawa *et al.*, 1980; Figure 3.10). Furthermore, probes of surface hydrophobicity also indicate a lower surface hydrophobicity for α_{s1}-CN (f25–199). Interestingly, the minor genetic variant α_{s1}-CN A, which is missing residues 14–26 (Swaisgood, 1982), behaves similarly to α_{s1}-CN (f25–199). For example α_{s1}-CN A has increased solubility in the presence of Ca^{2+} and a lower surface hydrophobicity. Thus, the self-association of α_{s1}-casein, and most likely its interaction with other caseins, appears to be facilitated by hydrophobic interaction of the N-terminal domain, possibly through a β-sheet interaction. Moreover, limited proteolysis of α_{s1}-casein yielding α_{s1}-CN (f25–199) and α_{s1}-CN (f1–23) has been correlated with a change in cheese rheology characterized by a loss of elasticity and curd strength (Creamer *et al.*, 1982).

At pH 4.0, limited hydrolysis of ovalbumin by pepsin results in the specific cleavage of His_{22} Ala_{23} giving a large domain, ovalbumin (f23–385) (Kitabatake *et al.*, 1988). This polypeptide is extremely stable so that proteolysis can be continued until the yield is essentially 100%, strongly suggesting that ovalbumin (f23–385) represents a compact domain of this protein. Of even more interest to the food scientist is the observation that, as a result of this limited proteolysis, transparent thermally induced gels are formed rather than the opaque gels normally obtained with ovalbumin (Kitabatake *et al.*, 1988). Hence, the functionality of ovalbumin (f23–385) is obviously considerably changed from that of the parent protein.

3.6 Preparation of a possible structural domain of β-lactoglobulin by limited proteolysis

Using immobilized trypsin, conditions for limited proteolysis of β-lactoglobulin that appear to result in the release of a structural domain were established (Chen and Swaisgood, 1991). Trypsin was covalently immobilized on succinamido-propyl porous glass using the sequential activation/immobilization procedure developed in our laboratory (Janolino and Swaisgood, 1982). Various conditions for limited proteolysis with immobilized trypsin were examined. A typical SDS-PAGE pattern for the hydrolysate obtained after several time periods of proteolysis at pH 8.0, 20 mM Tris buffer at 10°C is shown in Figure 3.11. A 10 min period of limited proteolysis appeared optimal for production of a 13 kD polypeptide. Fractionation of the limited digest by gel chromatography with Sephadex G-50 followed by BioGel P30 allowed purification of the 13 kD 'domain' (Figure 3.12). This preparation is not completely homogeneous, as suggested by Figures 3.11 and 3.12, and may represent polypeptides produced by cleavage at adjacent sites yielding 11–13 kD domains. Some preliminary results on several characteristics of this preparation have been obtained.

As indicated in Figures 3.11 and 3.12, the product of limited proteolysis of β-lactoglobulin elutes just prior to the native protein, suggesting a slightly expanded structure. Such observations could result from molecular expansion of a released domain or a loosening of structure in a peptide associated with the 13 kD domain by noncovalent interactions or disulfide bonds. A change in the surface hydrophobicity is indicated by increased retardation of the modified protein or domain with propyl-silica upon hydrophobic interaction chromatography. The amount of secondary structure was examined by CD spectroscopy. As indicated by the spectra given in Figure 3.13, the limited proteolysis product has a secondary structure very similar to that of the parent protein. Thus, most of the structure must be retained as would be expected if proteolysis is limited to a flexible loop or linker between structural domains. The stability of the structure was also examined and compared with that of the parent protein. Preliminary data on

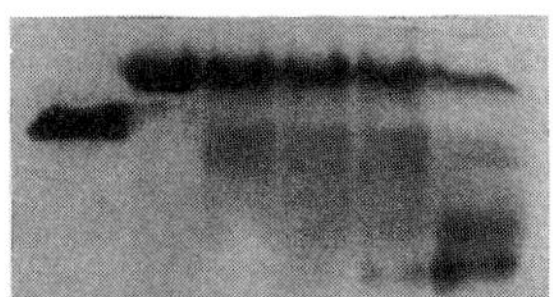

Figure 3.11 SDS-polyacrylamide gel electrophoresis patterns for digestion of β-lactoglobulin by immobilized trypsin. Digestion was performed in 20 mM Tris buffer, pH 8.0, at 10°C using 500 units of activity per gram β-lactoglobulin. Protein bands were visualized by silver-staining and represent lysozyme standard (14.3 kD) in lane 1, native β-lactoglobulin in lane 2, and hydrolysates in lanes 3, 4, 5 and 6 corresponding to 10 min, 20 min, 30 min, and 60 min, respectively.

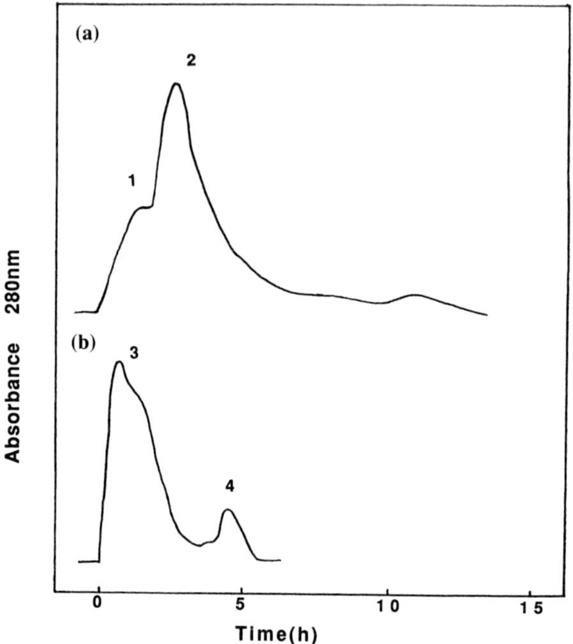

Figure 3.12 Fractionation of the immobilized trypsin digest of β-lactoglobulin by gel chromatography. (a) Chromatography of the digest on Sephadex G-50 (2.5 × 100 cm) in 50 mM glycine, 50 mM NaCl, pH 2.8 containing 0.02% sodium azide. Peak 1 represents 'domain' and peak 2 is native β-lactoglobulin. (b) Chromatography of peak 1 from Sephadex G-50 on Bio Gel P-30 (1 × 100 cm) in the same buffer used for Sephadex elution. Peak 3 represents the 'domain' and peak 4 corresponds to native β-lactoglobulin.

the structural stability in urea solutions indicate that the domain has a significantly lower stability, exhibiting a co-operative transition at a urea concentration roughly 1 M less than that of native β-lactoglobulin. Previously we have shown that native β-lactoglobulin binds specifically and with high affinity ($K_D \approx 10^{-8}$ M) to *trans*-retinal immobilized on porous glass (Jang and Swaisgood, 1990). Hence the ability of domains to bind to immobilized retinal was examined. Using bovine serum albumin (BSA) as a control to indicate any possible non-specific binding, purified domain and native β-lactoglobulin were incubated with immobilized retinal beads after which the supernatant was analyzed by SDS-PAGE. Results showed that both were completely removed from solution by adsorption whereas the BSA remained (Chen and Swaisgood, 1991).

The X-ray crystal structure of β-lactoglobulin (Papiz *et al.*, 1986; Monaco *et al.*, 1987) indicates a core structural motif consisting of an eight-stranded antiparallel β-barrel on the surface of which another β-strand and an α-helix is positioned. The centre of the β-barrel is clearly hydrophobic and is the retinol-binding pocket for the similarly structured retinol-binding proteins (Papiz *et al.*, 1986; Monaco *et al.*,

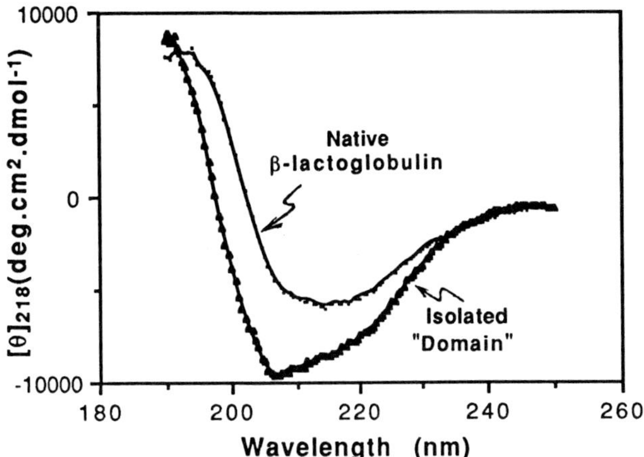

Figure 3.13 Circular dichroism spectra of isolated 'domain' compared with that of native β-lactoglobulin. These spectra were obtained in 50 mM glycine, 50 mM NaCl, pH 2.8, containing 0.02% sodium azide at 22–25°C.

1987). However, there also appears to be a hydrophobic surface pocket between the surface of the β-barrel and the α-helix. Crystallographic evidence suggests that this surface point is the binding site for retinol in β-lactoglobulin.

Examination of both primary (Swaisgood, 1982) and three-dimensional (Papiz *et al.*, 1986; Monaco *et al.*, 1987) structure suggests that Lys_{60} in the N-terminal region or Lys_{101} and/or Arg_{123} in the C-terminal region are likely candidates for limited trypsin cleavage to produce 11–13 kD polypeptides. Each of these residues occur on or adjacent to reverse turns or loops. The authors favour the choice of Arg_{123} as the most likely cleavage site since the other possibilities occur in the β-barrel motif, which may be more structurally stable. Furthermore, reduction of the isolated proteolysis product with dithiothreitol in physiological buffer yielded smaller domains as measured by size exclusion chromatography. Such conditions would be expected to affect only those peptides attached to the domain by disulfide bonds. Also, the apparently specific binding to immobilized retinal suggests that either the β-barrel motif contains the binding pocket as originally suggested (Papiz *et al.*, 1986) or that a sufficient portion of the C-terminal peptide remains attached by the disulfide bond to allow a strong interaction.

These preliminary studies of limited proteolysis of β-lactoglobulin demonstrate that large structured domains can be produced. Furthermore, the increased surface hydrophobicity and alteration of the structural stability indicate that dramatic changes in the functionality could be expected. Future studies will identify the site(s) of limited proteolysis, e.g. sequence analyses, and determine the effect of disulfide bond reduction on structure and binding to immobilized retinal. Several functional characteristics of the isolated domain will also be examined in anticipation that more-complete knowledge of the structure of a simple domain will significantly contribute to our understanding of the relationship between structure and function.

3.7 Potential design of protein functionality by limited transglutaminase-catalyzed cross-linking of domains

Relationships between the evolution of protein structure and the existence of structural and functional domains are increasingly being recognized (Neurath, 1986). Whether or not there is a direct relationship between exons and domain structure, it is apparent that structural motifs have been recruited for common structural functions and that domains with specific biological functions are shared by otherwise seemingly unrelated proteins. Thus, recombination of structural and functional domains under the selective pressure of evolution has led to multidomain proteins capable of the regulation required for the fine-tuning of biological processes.

Likewise, it would seem that recombination of domains could provide for the fine-tuning of the functionality of proteins in foods. Of course, for some proteins it would be possible to genetically recombine domains by construction of fusion proteins, provided that it is first understood which domains should be combined to give a specific functionality. However, there is perhaps an easier and more general enzymatic way for recombination of protein domains. Presuming that domains can be released from proteins by selective conditions of limited proteolysis, it may be possible to prepare proteins representing a recombination of domains by limited cross-linking of a selected mixture. Of course, the preparation would represent an essentially random recombination of the domains present in the mixture but the extent of cross-linking and the selection of domains in the mixture should provide opportunity for design of functionality.

The enzyme transglutaminase has been used to modify functionality by the cross-linking of whole proteins. For example, the cross-linking of β-lactoglobulin (Tanimoto and Kinsella, 1988) changed the viscosity, heat stability and gelling properties of that protein. Likewise, gelation of casein and soybean globulins is catalyzed by transglutaminase (Nio *et al.*, 1985) and the enzyme can be used to form cross-links between myosin and meat extenders such as casein and soy protein (Kurth and Rogers, 1984).

Pursuing the concepts developed above, investigations of immobilized transglutaminase as a potential bioreactor catalyst for limited cross-linking of proteins and/or protein domains have been initiated. To the authors' knowledge, this is the first reported study of covalently immobilized transglutaminase. Preliminary studies of the enzyme covalently attached directly to succinamidopropyl glass suggested that significant activity was not retained. Therefore, to improve the steric accessibility of the enzyme to large protein substrates, a protein spacer was incorporated between the matrix and the enzyme. α_{s1}-casein was covalently immobilized on glutaraldehyde-activated aminopropyl-glass, the concentration of amino groups increased by transglutaminase-catalyzed reaction of α_{s1}-casein-glass with poly-L-lysine, and finally, the poly-L-lysyl-α_{s1}-casein-glass was activated with glutaraldehyde and transglutaminase was added. These preparations exhibited activity with either β-lactoglobulin or α_{s1}-casein as substrates (Table 3.1). Comparison with the soluble enzyme indicated that immobilized enzyme retained roughly 11–13% of the catalytic efficiency (kcat/K_M); a rather typical value for immobilized enzymes exhibiting pore mass-transfer limitations and some inactivation due to immobilization. The data also show that α_{s1}-casein is much better than β-lactoglobulin as a substrate, with an approximate five-fold difference in catalytic efficiencies. From these results it is concluded that immobilized transglutaminase has potential as a biocatalyst for the cross-linking of protein domains.

Table 3.1 Transglutaminase-catalyzed cross-linking of proteins[a]

Enzyme	β-lactoglobulin			α_{s1}-casein		
	k_{CAT} (min^{-1})	K_M (mg/ml)	k_{CAT}/K_M (ml/mg.min^{-1})	k_{CAT} (min^{-1})	K_M (mg/ml)	k_{CAT}/K_M (ml/mg.min^{-1})
Soluble	5.00	0.61	8.2	5.31	0.14	37.9
Immobilized	0.27	0.29	0.9	1.86	0.37	5.0

[a]Activities were measured using a fluorescent o-phthalaldehyde technique for measuring the disappearance of primary amino groups. Enzyme catalysis was observed in 25 mM imidazole buffer, at pH 7.5, containing 5 mM $CaCl_2$ and 10 mM DTT at 22–25°C.

3.8 Conclusions and future directions

It is now recognized that most proteins are assemblies of multiple functional and/or structural domains. Genetic recombination of these units of protein structure under selective evolutionary pressure allows development of new biological functions and the fine-tuning of metabolic regulation. By analogy to the evolutionary design of biological function, it is proposed that the functionality of proteins in foods could be designed by recombination of domains.

Domains are separated by peptide sequences having greater flexibility, hence these connections are more susceptible to proteolysis. Consequently, domains can be liberated from their parent proteins by limited proteolysis. Using the advantages of immobilizing proteinases, it may be possible to obtain bioreactors that are capable of preparing large quantities of protein domains. The availability of such domains will permit the examination of these simpler units of protein structure for the relationship between structure and functionality, which should greatly enhance our understanding of this fundamental problem.

It may also be possible, for example using immobilized transglutaminase, to cross-link specific mixtures of domains. By purification of the products, such limited cross-linking would allow the functionality of various combinations of domains to be explored, e.g. the characteristics of multiple domains having different thermal stabilities, polarities, or surface hydrophobicities.

Finally, selection of domains in mixtures for limited and random cross-linking by immobilized transglutaminase bioreactors may permit sufficient design of functionality for large-scale bioprocessing.

References

Betton, J.-M., Desmadril, M. and Yon, J.M. (1989) Detection of intermediates in the unfolding transition of phosphoglycerate kinase using limited proteolysis. *Biochem.* **28**:5421–28.

Burgess, A.W., Weinstein, L.I., Gabel, D. and Scheraga, H.A. (1975) Immobilized carboxypeptidase A as a probe for studying the thermally induced unfolding of bovine pancreatic ribonuclease. *Biochem.* **14**:197–200.

Chaplin, L.C., Clark, D.C. and Smith, L.J. (1988) The secondary structure of peptides derived from caseins: circular dichroism study. *Biochim. Biophys. Acta* **956**:162–72.

Chen, S.X. and Swaisgood, H.E. (1991) Characteristics of a structural domain prepared from β-lactoglobulin by limited proteolysis with immobilized trypsin. *J. Dairy Sci.* **74** (*Suppl. 1*): 102.

Church, F.C., Catignani, G.L. and Swaisgood, H.E. (1982a) Urea denaturation of the immobilized proteases of *Streptomyces griseus* (Pronase). *Enzyme Microb. Technol.* **4**:313–16.

Church, F.C., Catignani, G.L. and Swaisgood, H.E. (1982b) Use of immobilized *Streptomyces griseus* proteases (Pronase) as a probe of structural transitions of lysozyme, β-lactoglobulin and casein. *Enzyme Microb. Technol.* **4**:317–21.

Creamer, L.K., Zoerb, H.F., Olson, N.F. and Richardson, T. (1982) Surface hydrophobicity of α_{s1}-I, α_{s1}-casein A and B and its implications in cheese structure. *J. Dairy Sci.* **65**:902–6.

Di Pietro, A., Godinot, C. and Gautheron, D.C. (1983) Use of trypsin to monitor conformational changes of mitochondrial adenosinetriphosphatase induced by nucleotides and phosphate. *Biochem.* **22**:785–92.

Eigel, W.N. (1977) Formation of γ_1-A^2, γ_2-A^2 and γ_3-A caseins by *in vitro* proteolysis of β-casein A^2 with bovine plasmin. *Int. J. Biochem.* **8**:187.

Fischetti, V.A., Parry, D.A.D., Trus, B.L., Hollingshead, S.K., Scott, J.R. and Manjula, B.N. (1988) Conformational characteristics of the complete sequence of group A streptococcal M6 protein. *Proteins: Struct. Funct. Genet.* **3**:60–9.

Fontana, A., Fassina, G., Vita, C., Dalzoppo, D., Zamai, M. and Zambonin, M. (1986) Correlation between sites of limited proteolysis and segmental mobility in thermolysin. *Biochem.* **25**:1847–51.

Girma, J.-P., Chopek, M.W., Titani, K. and Davie, E.W. (1986) Limited proteolysis of human von Willebrand factor by *Staphlococcus aureus* V-8 protease: Isolation and partial characterization of a platelet-binding domain. *Biochem.* **25**:3156–63.

Hubbard, M.J. and Klee, C.B. (1989) Functional domain structure of calcineurin A: Mapping by limited proteolysis. *Biochem.* **28**: 1868–74.

Jang, H.D. and Swaisgood, H.E. (1990) Analysis of ligand binding and β-lactoglobulin denaturation by chromatography on immobilized *trans*-retinal. *J. Dairy Sci.* **73**: 206–74.

Janolino, V.G. and Swaisgood, H.E. (1982) Analysis and optimization of methods using water-soluble carbodiimide for immobilization of biochemicals to porous glass. *Biotechnol. Bioeng.* **24**:1069–80.

Kaminogawa, S., Yamauchi, K. and Yoon, C.-H. (1980) Calcium insensitivity and other properties of α_{s1}-casein. *J. Dairy Sci.* **63**:223–7.

Khandke, K.M., Fairwell, T., Acharya, A.S. and Manjula, B.N. (1990) Domain structure and molecular flexibility of streptococcal M protein *in situ* probed by limited proteolysis. *J. Protein Chem.* **9**:511–21.

Kitabatake, N., Indo, K. and Doi, E. (1988) Limited proteolysis of ovalbumin by pepsin. *J. Agric. Food Chem.* **36**:417–20.

Kohn, J. and Wilchek, M. (1982) Mechanism of activation of Sepharose and Sephadex by cyanogen bromide. *Enzyme Microb. Technol.* **4**:161–63.

Kumosinski, T.F., Brown, E.M. and Farrell, H.M. Jr. (1991) Three-dimensional molecular modeling of bovine caseins: α_{s1}-caseins. *J. Dairy Sci.* **74**:2889–95.

Kurth, L. and Rogers, P.J. (1984) Transglutaminase catalyzed cross-linking of myosin to soya protein, casein and gluten. *J. Food Sci.* **49**:573–76.

Linderstrom-Lang, K. (1952) The initial stages in the breakdown of proteins by enzymes. *Lane Medical Lectures Stanford Univ. Ser. Med. Sci.* **6**:53.

Marshall, M. and Fahlen, L.A. (1988) Proteolysis as a probe of ligand-associated conformational changes in rat carbamyl phosphate synthetase I. *Arch. Biochem. Biophys.* **262**:455–70.

Matthyssens, G. Simons, G. and Kanarek, L. (1972) Study of the thermal-denaturation mechanism of hen egg-white lysozyme through proteolytic digestion. *Eur. J. Biochem.* **26**:449–54.

McClintock, D.K. and Markus, G. (1968) Conformational changes in aspartate transcarbamylase. *J. Biol. Chem.* **243**:2855–62.

McLachlan, A.D. (1984) Structural implications of the myosin amino-acid sequence. *Ann. Rev. Biophys. Bioeng.* **13**:167–89.

Mihalyi, E. (1978) *Application of Proteolytic Enzymes to Protein Structure Studies*, Vol. 1, CRC Press, Boca Raton, Florida.

Monaco, H.L., Zanotti, G., Spadon, P., Bolognesi, M., Sawyer, L. and Eliopoulos, E.E. (1987) Crystal structure of the trigonal form of bovine beta-lactoglobulin and of its complex with retinol at 2.5 Å resolution. *J. Mol. Biol.* **197**:695–706.

Neurath, H. (1986) Limited proteolysis, domains, and the evolution of protein structure. *Chem. Scr.* **26B**:221–29.

Nio, N., Motoki, M. and Takinami, K. (1985) Gelation of casein and soybean globulins by transglutaminase. *Agric. Biol. Chem.* **49**:2283–86.

Papiz, M.Z., Sawyer, L., Eliopoulos, E.E., North, A.C.T., Findlay, J.B.C, Sivaprasadarao, R., Jones, T.A., Newcomer, M.E. and Kraulis, P.J. (1986) The structure of β-lactoglobulin and its similarity to plasma retinol-binding protein. *Nature* **324**:383–85.

Porter, D.H., Swaisgood, H.E. and Catignani, G.L. (1984) Characterization of an immobilized digestive enzyme system for determination of protein digestibility. *J. Agric. Food Chem.* **32**:334–39.

Rawn, J.R. (1989) *Biochemistry*. Neil Patterson Publishers, Carolina Biological Supply, Burlington.

Rupley, J.A. (1967) Susceptibility to attack by proteolytic enzymes. *Method Enzymol.* **11**:905–17.

Shimizu, M., Lee, S.W., Kaminogawa, S. and Yamauchi, K. (1986) Functional properties of a peptide of 23 residues purified from the peptic hydrolyzate of α_{s1}-casein: changes in the emulsifying activity during purification of the peptide. *J. Food Sci.* **51**:1248–52.

Swaisgood, H.E. (1989) Structural changes in milk proteins, in *Milk Proteins* (eds. C.A. Barth and E. Schlimme) Steinkopf Verlag Darmstadt, New York, pp. 192–210.

Swaisgood, H.E. (1982) Chemistry of milk protein, in *Developments in Dairy Chemistry*, Vol. 1, (ed. P.F. Fox) Applied Science Publishers, London, pp. 1–59.

Swaisgood, H.E., Horton, H.R. and Mosbach, K. (1976) Covalently bound glutamate dehydrogenase for studies of subunit association and allosteric regulation. *Methods Enzymol.* **44**:504–15.

Swaisgood, H.E. and Catignani, G.L. (1987) Use of immobilized proteinases and peptidases to study structural changes in proteins. *Methods Enzymol* **135**:596–604.

Swaisgood, H.E. and Horton, H.R. (1989) Immobilized enzymes as processing aids or analytical tools, in *Biocatalysis in Agricultural Biotechnology* (eds J.R. Whitaker and P.E. Sonnet) ACS Symposium Series 389, American Chemical Society, Washington DC. pp. 242–261.

Tanimoto, S.-Y. and Kinsella, J.E. (1988) Enzymatic modification of proteins: Effects of transglutaminase cross-linking on some physical properties of β-lactoglobulin. *J. Agric. Food Chem.* **36**:281–85.

Ueno, H. and Harrington, W.F. (1984) An enzyme-probe method to detect structural changes in the myosin rod. *J. Mol. Biol.* **173**:35–61.

Ueno, H. (1984) Local structural changes in tropomyosin detected by a trypsin-probe method. *Biochem.* **23**:4791–4798.

4 Control of polyphenol oxidase activity using a catalytic mechanism

D. OSUGA, A. VAN DER SCHAAF AND J.R. WHITAKER

Abstract

Polyphenol oxidase is one of the most deteriorative of enzymes, especially in tropical fruits, yet it is essential for color of brown tea, cocoa, coffee, raisins, some figs and prunes and plant protection. It is responsible for the unwanted black-spot formation in shrimp but is important for pigmentation of human skin. In this chapter the mechanism of action of polyphenol oxidase is discussed, including the reactions catalyzed, the kinetics with respect to the two substrates, O_2 and phenol, substrate specificity and the intermediates in the reaction. Differences between monophenolase and diphenolase activities are shown mechanistically. Complete amino-acid sequences are available for polyphenol oxidases from humans and mice, *Neurospora crassa* (fungus), and *Streptomyces glaucescens* (fungus) and *S. antibioticus* (fungus). There is 86% strict homology in amino-acid sequence between *S. glaucescens* and *S. antibioticus* but only 24% between the *Streptomyces* enzymes and that from *N. crassa*. The human and mouse enzymes are 43% homologous; they are much larger than the fungal enzymes and have little homology with the fungal enzymes, except in the active-site histidine residues. There is also much homology between the polyphenol oxidases and hemocyanins around the active-site histidine reactions. Several methods of controlling polyphenol oxidase utilize pH, O_2 exclusion, heating, ascorbic acid, sodium bisulfite, thiol compounds, k_{cat} inactivation, competitive inhibitors and removal of phenols.

4.1 Introduction

Polyphenol oxidase (EC 1.14.18.1; also known as: monophenol; dihydroxyphenylalanine; oxygen oxidoreductase; other trivial names including tyrosinase, phenolase, catechol oxidase, polyphenol oxidase, monophenol oxidase; (Enzyme Nomenclature, 1979)) is found in most plant tissues, some animal tissues including humans, shrimp and insects, and in some fungi. The function of polyphenol oxidase is best understood in humans where this enzyme is responsible for skin pigmentation, including freckles. Tyrosine is the primary substrate in humans, justifying the trivial name tyrosinase. Its *in vivo* role in formation of dihydroxy-phenylalanine (DOPA) is not clear. The physiological role(s) in higher plants and fungi is also not entirely certain. Szent-Györgyi and Vietorisz (1931) suggested that polyphenol oxidase is a protective enzyme since it is responsible for forming a 'scab' of insoluble melanin over wounds in plants, sealing in the sap and sealing out insects and microorganisms.

Polyphenol oxidase is undoubtedly the most important deteriorative enzyme in fruits, especially tropical fruits, where up to one-half of the fruit may be lost. Bruises, cuts and other mechanical damage that allow O_2 penetration lead to rapid browning due to melanin formation. Not only is the color unacceptable to the consumer but there is an off-taste and a loss of nutritional quality caused primarily by reaction of the enzymatic product, *o*-benzoquinone, with the ϵ-amino group of lysine in proteins. Peaches, apricots, apples and grapes also brown, as do some vegetables, especially Irish potatoes, some lettuce and other leafy vegetables. Black-spot development in shrimp is a major problem. However, in tea, coffee, cocoa, prunes, black raisins, black figs (e.g. Black Mission) and some other fruits, browning due to the action of polyphenol oxidase is desirable. Recent investigations have shown that polyphenol oxidase products (melanins) are excellent sun blockers when applied to the skin, and some of the soluble pigments have potential as food colorants (V. Kahn and J.R. Whitaker, 1994, unpublished data). Absence of melanin biosynthesis results in oculo-cutaneous albinism in humans (Witkop, 1984) and an over-production of melanin by melanocytes may lead to toxic intermediates (Hochstein and Cohen, 1963).

The purpose of this chapter is to explore: (i) the mechanism of action of polyphenol oxidases; (ii) the homologies in amino-acid sequence among several polyphenol oxidases, including fungal, bacterial, human and mouse; and (iii) methods for the control of polyphenol oxidase activity.

4.2 Mechanism of action of polyphenol oxidase

4.2.1 Reactions catalyzed

Polyphenol oxidases can catalyze two quite different types of reactions, as shown in equations 4.1 and 4.2.

$$\text{p--Cresol} + O_2 + BH_2 \longrightarrow \text{4--Methylcatechol} + B + H_2O \qquad (4.1)$$

$$2 \quad \text{Catechol} \quad + O_2 \longrightarrow \quad \text{o–Benzoquinone} \quad + 2H_2O \tag{4.2}$$

Catechol   o–Benzoquinone

There is disagreement as to whether all polyphenol oxidases can catalyze equation 4.1. Peach (Wong *et al.*, 1971a,b) and pear (Rivas and Whitaker, 1973) polyphenol oxidases do not oxidize monohydroxy phenols, such as *p*-cresol, within several hours of reaction time. Hydroxylation of monophenols is generally slower than oxidation of *o*-diphenols for polyphenol oxidases that catalyze both reactions (Lerch and Ettlinger, 1972). Mushroom and human skin polyphenol oxidases catalyze equation 4.1 readily (Bouchilloux *et al.*, 1963). In equation 4.1, BH_2 is an *o*-dihydroxyphenol required as a co-factor, as shown in equation 4.3 (see page 76). Some, if not all, polyphenol oxidases that hydroxylate monohydroxy phenols (e.g. those from mushroom and human skin), can slowly form BH_2, thereby activating the enzymes; these enzymes show hysteresis in that initial rate of product formation accelerates with time. Whether peach and pear polyphenol oxidases lack the ability to form catalytic amounts of BH_2, or lack the ability to insert oxygen at the 3-position (e.g. *p*-cresol) is unclear. Use of 3-tritiated *p*-cresol, with or without catalytic amounts of catechol ($K_d \simeq 0.1 \ \mu M$), would probably resolve this question.

All polyphenol oxidases discovered so far have the ability to convert *o*-dihydroxyphenols to *o*-benzoquinones using the second substrate, O_2 (equation 4.2). Further polymerization of *o*-benzoquinones to melanins also requires O_2 but the reactions are generally nonenzymic. Polymerization to melanins involves Michael addition to the benzene ring of the very reactive *o*-benzoquinone. Some polymerization products are soluble, some are insoluble. Colors range from yellow, brown, red, green and blue depending on the extent of polymerization and nature of the chromophoric groups.

The 4-methyl group in *p*-cresol and 4-methylcatechol can be replaced with such chemical moieties as H, longer alkyl chains, COOH, F and NO_2. 1,3- and 1,4-dihydroxyphenols and 1,3,5-trihydroxyphenol are not substrates for polyphenol oxidases, but may be inhibitors or form Michael addition compounds with enzymatically produced *o*-benzoquinones. Another enzyme, laccase, hydroxylates monophenols at the *p*-position and subsequently forms *p*-benzoquinones (Mason, 1965); it is frequently confused with polyphenol oxidase since Enzyme Nomenclature (1979) also assigns it the number EC 1.14.18.1.

Both reactions 1 and 2 (equations 4.1 and 4.2) require O_2, as demonstrated using $^{18}O_2$. The K_m for O_2 is 0.11 mM for pear polyphenol oxidase (Rivas and

Whitaker, 1973) and 0.06 to 0.30 mM for *Streptomyces glaucescens* polyphenol oxidase (Lerch and Ettlinger, 1972). Since solubility of O_2 in aqueous buffers near neutrality is 0.24 mM, the enzymes above are not saturated with O_2 under atmospheric conditions.

4.2.2 Kinetic mechanism

Kinetically, polyphenol oxidases follow a Bi Bi Ordered Sequential Mechanism (Figures 4.1 and 4.2) (Cleland, 1963a–c; Lerch and Ettlinger, 1972; Rivas and Whitaker, 1973). The order in which the two products, H_2O and benzoquinone, are released from the enzyme is unknown. Rates of reactions can be followed spectrophotometrically at 280 nm (see equation 4.1) or 385–475 nm (see equation 4.2), depending on the substrate, or polarographically (O_2 uptake). Three atoms of oxygen are used enzymically (if both equations 4.1 and 4.2 apply) per phenol substrate and up to three atoms of oxygen are used nonenzymically to form melanin, depending on the complexity of melanin formed. The spectrophotometric and polarographic methods give the same results using initial

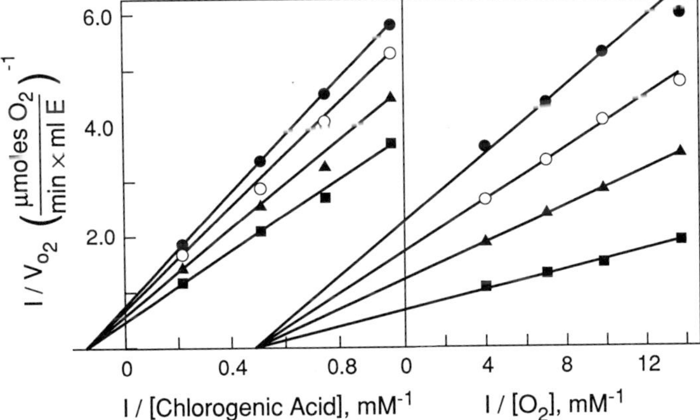

Figure 4.1 Effect of oxygen and chlorogenic acid concentrations on initial velocity of pear enzyme B-catalyzed reactions. The reactions were followed with an oxygen electrode at pH 4.0 and 30.0°C. (See Rivas and Whitaker, 1973, for experimental details.)

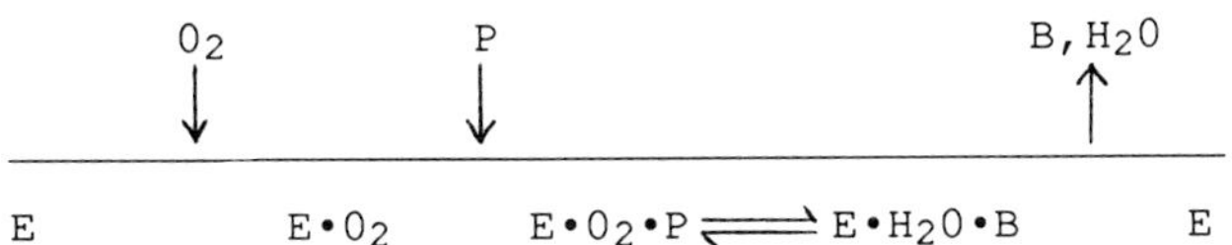

Figure 4.2 Kinetic mechanism for polyphenol oxidase, according to the Cleland nomenclature (1967a–c). E is enzyme, P is a phenol, B is a benzoquinone, E · O$_2$ is a binary enzyme · O$_2$ adsorptive complex, E · O$_2$ · P is a ternary enzyme · O$_2$ · phenol adsorptive complex and E · H$_2$O · B is a ternary enzyme · H$_2$O · benzoquinone adsorptive complex.

velocities, v_o, (<5% conversion of substrate to product), but not later when the nonenzymic reactions become important.

Reaction 1 (equation 4.1) cannot be studied independently of reaction 2 (equation 4.2) but can be monitored separately using *o*-tritiated monophenol or absorbance at 280 nm.

4.2.3 Substrate specificity

The polyphenol oxidases have broad substrate specificity. A detailed study of the substrate specificity of *S. glaucescens* is shown in Tables 4.1 and 4.2 (Lerch and Ettlinger, 1972). Based on the specificity coefficient k_{cat}/K_m, the best monophenol substrate for this enzyme is N-acetyl-L-tyrosine ethyl ester, and the worst is L-tyrosine. The best *o*-diphenol substrate is *t*-butylcatechol, which is 295 times better than 3,4-dihydroxy-D-phenylalanine, the worst substrate. Note that 3,4-dihydroxy-L-phenylalanine is 20 times a better substrate than is 3,4-dihydroxy-D-phenylalanine.

Table 4.1   Michaelis-Menten (K_m), catalytic (k_{cat}), k_{cat}/K_m constants and lag-time (t) of different monophenol substrates[a]

Substrate	K_m mM	k_{cat} s^{-1}	t min	k_{cat}/K_m $s^{-1}\,mM^{-1}$
L-Tyrosine	0.410	13.2	5.00	32.2
L-Tyrosine methyl ester	1.667	235	0.85	141
L-Tyrosine ethyl ester	0.945	223	0.55	236
L-Tyrosinamide	1.540	172	0.55	112
L-Tyrosine hydrazide	2.960	151	1.75	51.0
L-Tyrosyl-L-leucine	1.815	357	0.95	197
L-Tyrosyl-glycyl-glycine	3.058	310	1.30	101
N-Formyl-L-tyrosine	0.689	104	5.00	151
N-Acetyl-L-tyrosine	0.553	182	3.30	329
N-Chloroacetyl-L-tyrosine	0.287	249	2.20	868
N-Trifluoroacetyl-L-tyrosine	0.467	330	2.80	707
Glycyl-L-tyrosine	5.341	412	4.20	77.1
L-Leucyl-L-tyrosine	2.225	152	9.70	68.3
N-Acetyl-L-tyrosine ethyl ester	0.268	243	1.15	907
N-Chloroacetyl-L-tyrosine ethyl ester	0.052	158	1.10	3038
N-Trifluoroacetyl-L-tyrosine ethyl ester	0.045	104	1.05	2311
N-Acetyl-L-tyrosinamide	0.225	227	0.70	1009
N-Acetyl-L-tyrosine hydrazide	0.247	376	0.32	1522
p-Cresol	0.242	11.6	3.76	47.9
p-Hydroxyphenylacetic acid	1.164	48.8	16.00	41.9
p-Hydroxyphenylpropionic acid	0.488	134	4.30	275
L-Tyrosine methyl ester (α)[b]	0.462	148	0.70	320
L-Tyrosine methyl ester (γ)[b]	0.262	129	0.70	492

[a]Adapted from Lerch and Ettlinger, 1972. Activity was measured polarographically at 30°C in 0.1 M phosphate buffer, pH 7.0; [O_2] was 0.24 mM (air saturated buffer). *Streptomyces glaucescens* polyphenol oxidase was used.
[b]Kinetic constants for α and γ isozymes of mushroom tyrosinase (kinetic constants based on the subunit molecular weight of 32 400 Daltons) (Jolley *et al.*, 1969).

Table 4.3 provides a measure of the relative 'goodness' of several monophenol/*o*-diphenol pairs. In all cases the paired *o*-diphenol is the better substrate by a factor of 1.77 to 25.7 for 3,4-dihydroxy-L-phenylalanine/L-tyrosine methyl ester and homocatechol/*p*-cresol, respectively. Also included are some comparison data for mushroom polyphenol oxidase isozymes α and γ (Tables 4.1–4.3). Substrate specificity data for several other polyphenol oxidases are available in the literature (Yasunobu, 1959; Pomerantz, 1963; Sussman, 1961; Duckworth and Coleman, 1970) but they are more difficult to interpret.

Table 4.2 Michaelis-Menten (K_m), catalytic (k_{cat}) and k_{cat}/K_m constants for different *o*-diphenol substrates[a]

Substrate	K_m	k_{cat}	k_{cat}/K_m
	mM	s^{-1}	$s^{-1} mM^{-1}$
3,4-Dihydroxy-L-phenylalanine	5.774	1445	250
3,4-Dihydroxy-DL-phenylalanine	12.558	1542	123
3,4-Dihydroxy-D-phenylalanine	15.020	184	12.3
Catechol	4.508	395	87.6
Homocatechol	1.906	2340	1230
t-Butylcatechol	0.915	3320	3630
3,4-Dihydroxybenzoic acid	1.023	8.2	8.01
3,4-Dihydroxyphenylacetic acid	1.169	142	121
3,4-Dihydroxyphenylpropionic acid	1.055	1830	1734
Caffeic acid	0.372	476	1280
Chlorogenic acid	0.748	785	1050
3-Hydroxytyramine hydrochloride	11.900	875	73.5
3,4-Dihydroxy-L-phenylalanine, α [b]	0.375	414	1100
3,4-Dihydroxy-L-phenylalanine, γ [b]	0.400	670	1680

[a]Adapted from Lerch and Ettlinger, 1972. Activity was measured polarographically at 30°C in 0.1 M phosphate buffer, pH 7.0, using *Streptomyces glaucescens* polyphenol oxidase; [O_2] was 0.24 mM (air saturated buffer).
[b]Kinetic constants for α and γ isozymes of mushroom tyrosinase (kinetic constants based on the subunit molecular weight of 32 400 Daltons) (Jolley *et al.*, 1969).

Table 4.3 Ratios of k_{cat}/K_m and k_{cat}/k_{cat}, in parentheses, for some monophenol/*o*-diphenol substrate pairs[a]

Substrate pairs		(k_{cat}/K_m) diphenol/
Monophenol	Diphenol	(k_{cat}/K_m) monophenol
L-Tyrosine	3,4-Dihydroxy-L-phenylalanine	7.81 (110)[c]
p-Cresol	Homocatechol	25.7 (222)
p-Hydroxyphenylpropionic acid	3,4-Dihydroxyphenylpropionic acid	6.31 (13.6)
p-Hydroxyphenylacetic acid	3,4-Dihydroxyphenylacetic acid	2.89 (2.89)
L-Tyrosine methyl ester	3,4-Dihydroxy-L-phenylalanine	1.77 (6.17)
L-Tyrosine methyl ester (α)[b]	3,4-Dihydroxy-L-phenylalanine (α)[b]	3.44 (2.62)
L-Tyrosine methyl ester (γ)[b]	3,4-Dihydroxy-L-phenylalanine (γ)[b]	3.41 (5.20)

[a]Calculated from data in Tables 4.1 and 4.2.
[b]Based on kinetic constants for α- and γ-isozymes of mushroom polyphenol oxidase, using a subunit molecular weight of 32 400 Daltons (Jolley *et al.*, 1969).
[c]k_{cat}, diphenol/k_{cat}, monophenol from Tables 4.1 and 4.2.

4.2.4 Mechanism of action

So far as is known, all polyphenol oxidases are copper-containing proteins. Two copper atoms are required per active site, as shown in Figure 4.3 for *Neurospora* spp. polyphenol oxidase. The two coppers (oxidation state I or II) are probably bound to the imidazole side chains of His_{96}, His_{104}, His_{187}, His_{193}, His_{281} and His_{306} and R, which may be tyrosine (Huber *et al.*, 1985; numbering is that used in Figure 4.4).

The proposed mechanism for *Neurospora* spp. polyphenol oxidase is shown in Figure 4.5 (Lerch, 1983). Owing to the inherent differences in chemistry between the oxygenation reaction (equation 4.1) and the dehydrogenation reaction (equation 4.2) the two mechanistic reactions are presented separately but linked by a common oxy-polyphenol oxidase compound (Figure 4.5). Possible intermediates in the diphenol oxidation pathway are illustrated in Figure 4.5a. As shown, catechol replaces the two water molecules bound to the two Cu(II) of oxy-polyphenol oxidase to form the catechol·enzyme complex. The catechol is oxidized to benzoquinone, leaving the enzyme in the met form. To complete the cycle, another catechol molecule serves as a reductant to form the deoxy-polyphenol oxidase (Cu(I)) which can then combine with O_2 to reform the oxy-polyphenol oxidase. The resting enzyme is thought to be in the Cu(II) form (met-polyphenol oxidase).

For a monophenol (Figure 4.5b) BH_2 is required to reduce Cu(II) to Cu(I) (deoxy-polyphenol oxidase), which then adds O_2 across the two Cu(I)s to form the oxy-polyphenol oxidase intermediate. The O–O bond distance has been shown to be similar to that in hydrogen peroxide (Jolly *et al.*, 1974). Hydrogen peroxide can add to met-polyphenol oxidase to give oxy-polyphenol oxidase directly (Winkler *et al.*, 1981; Wilcox *et al.*, 1985). Azide and mimosine, inhibitors of polyphenol oxidase, can displace 'peroxide' (Himmelwright *et al.*, 1980). The monophenol (MP) adds to the oxy-polyphenol oxidase, giving the ternary enzyme·O_2·MP complexes, which are then oxidized to the dihydroxyphenol·enzyme and subsequently to the *o*-benzoquinone·enzyme. The *o*-benzoquinone is released from the enzyme, leaving the enzyme in the Cu(I) state, ready to react with O_2 for another cycle. The rearrangement of the binuclear site to a trigonal bipyramidal intermediate (shown in brackets in Figure 4.5b), as generally observed with square planar and tetragonal associative substitution chemistry, is a critical step in the pathway. This rearrangement is expected to result in a deprotonated peroxide co-ordinated to only one copper. The resulting

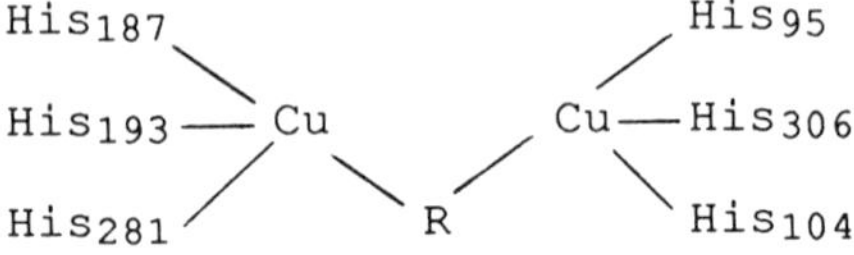

Figure 4.3 Proposed co-ordination of copper to six histidine residues and R (possibly tyrosine residue) of *Neurospora crassa* polyphenol oxidase (from Huber *et al.*, 1985).

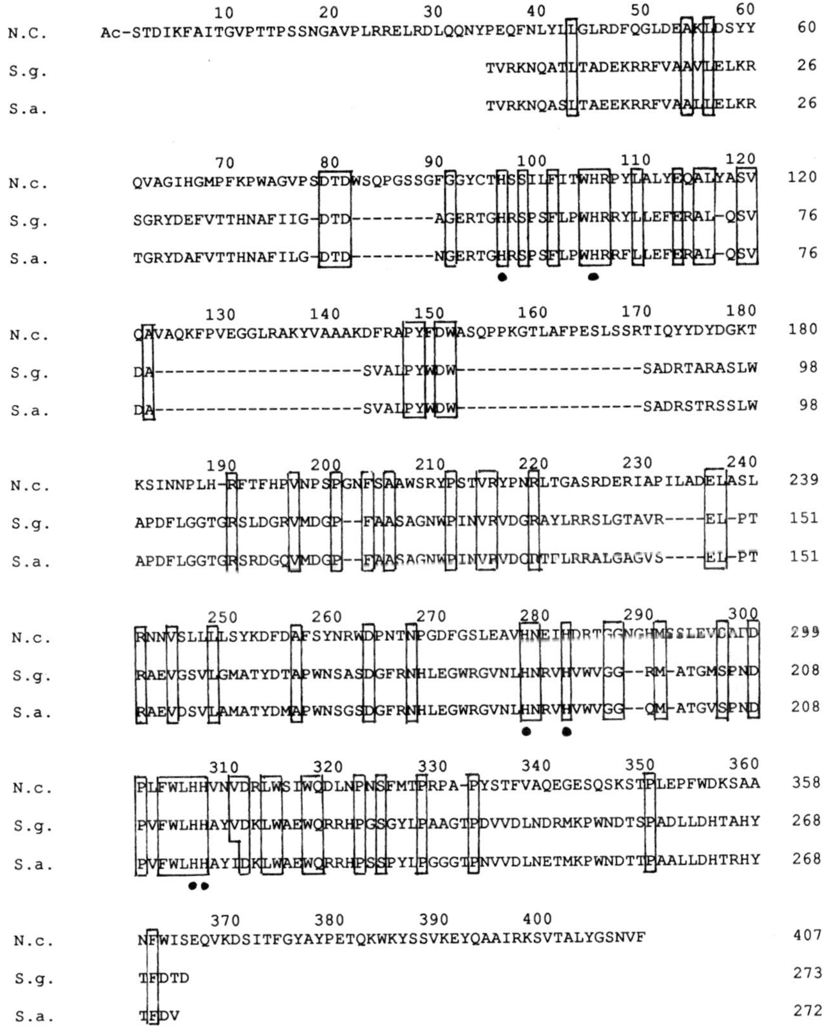

Figure 4.4 Primary sequences of *Neurospora crassa*, (N.c.) *Streptomyces glaucescens* (S.g.) and *S. antibioticus* (S.a.) polyphenol oxidases. The amino-acid sequences are according to Lerch (1978; *N. crassa*), Huber *et al.* (1985; *S. glaucescens*) and Bernan *et al.* (1985; *S. antibioticus*). The numbering system is that of *N. crassa* polyphenol oxidase. Identical sequences are boxed; a dash indicates a space left for alignment purposes.

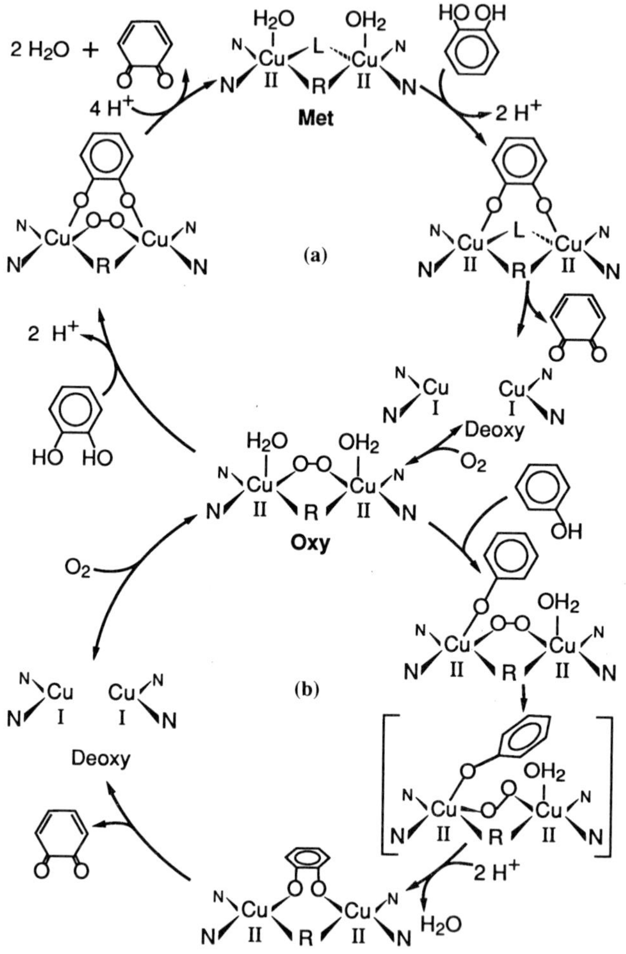

Figure 4.5 Proposed kinetic scheme of the diphenol (a) and monophenol (b) activities of polyphenol oxidase (from Lerch, 1983, with permission).

peroxide would be extremely reactive, easily able to hydroxylate the substrate in the *o*-position to the existing hydroxyl group.

4.3 Primary structures of polyphenol oxidases

The primary structures of at least seven polyphenol oxidases are known. These are the polyphenol oxidases from *Neurospora crassa* (Lerch, 1978, 1982), *S. glaucescens* (Huber *et al.*, 1985), *S. antibioticus* (Bernan *et al.*, 1985), mouse (Müller *et al.*, 1988), human (Kwon *et al.*, 1987), the protein encoded by pMT4 (Shibahara *et al.*, 1986) tomato (Shahar *et al.*, 1992) and potato (Hunt *et al.*,

1993). Surprisingly, even though the mushroom polyphenol oxidases have received the most attention, the amino-acid sequences have not been determined. The mushroom polyphenol oxidases are tetramers of 128 000 Daltons composed of four subunits of approximately 32 400 Daltons each (Jolley *et al.*, 1969). The nature and size of the subunits are still debated, as well as the relationship among the three isozymes α,β and γ. In contrast *S. glaucescens* polyphenol oxidase is a single polypeptide of 30 900 Daltons (273 amino acids), with no isozymes (Lerch and Ettlinger, 1972), while *N. crassa* polyphenol oxidase is a single polypeptide of 46 000 Daltons (407 amino acids) with several isozymes (Fling *et al.*, 1963).

The primary sequences of *N. crassa*, *S. glaucescens* and *S. antibioticus* polyphenol oxidases are shown in Figure 4.2 (Lerch, 1982; Bernan *et al.*, 1985). *N. crassa* polyphenol oxidase is much larger than *S. glaucescens* and *S. antibioticus* polyphenol oxidases, with molecular weights of 46 000, 30 000 and 30 000 Daltons, respectively. The overall sequence homology is 24% between the two groups of enzymes. There is little sequence homology at the N– (35–91) and C– (318–364) terminal ends (4.5 and 6.0%, respectively). There are two sequences at residues 92–121 and 277–317 with approximately 50% homology and a sequence in the middle (122–276) with approximately 14% sequence homology. The sequence 92–121 contains two histidine residues and the sequence 277–317 contains four histidine residues in identical positions in the three enzymes. These six histidine residues, with surrounding homology, are most likely the ligand sites for the two copper atoms in the binuclear active site. There is very high amino-acid sequence homology (86%, with the rest generally conservative changes) between *S. glaucescens* and *S. antibioticus* polyphenol oxidases.

The amino-acid sequences of human and mouse polyphenol oxidases are shown in Figure 4.6 (Shibahara *et al.*, 1986; Kwon *et al.*, 1987). The human and mouse polyphenol oxidases have molecular weights of 62 610 and 58 000 Daltons, respectively, which is considerably larger than the fungal enzymes. There is 43% conserved amino-acid sequence homology between the two inhibitors. Fifteen of the 16 1/2-cystine residues of mouse polyphenol oxidase align with 15 of the 17 1/2-cystine residues of human polyphenol oxidase. Among the three fungal polyphenol oxidases, only *N. crassa* contains a single 1/2-cystine residue. The human enzyme contains 13 histidine residues, while the mouse enzyme contains 16 residues. Seven of the histidine residues appear to be conserved. *N. crassa*, *S. glaucescens* and *S. antibioticus* polyphenol oxidases contain 10, 12 and 12 histidine residues, respectively. Six of the histidine residues are at identical positions in all three enzymes. Of the 12 histidine residues, 11 are in identical positions in the *S. glaucescens* and *S. antibioticus* enzymes.

Direct and indirect evidence indicates that most of the strictly conserved histidines are in the active site. Hemocyanins and polyphenol oxidases have several common features. Both bind O_2 but hemocyanins do not carry out catalysis. Both contain a binuclear copper atom system, as indicated by complex formation, and these are similar (Solomon, 1981; Lerch, 1981), especially in the different functional states (oxy-, deoxy- and met- forms; Figure 4.5). The binuclear copper atom site in polyphenol oxidase is much more accessible to a second compound, such as phenol, azide or mimosine (Winkler *et al.*, 1981) than the site in hemocyanin, permitting polyphenol oxidase to be an enzyme and hemocyanin an O_2 carrier protein. Remarkedly, the six different polyphenol

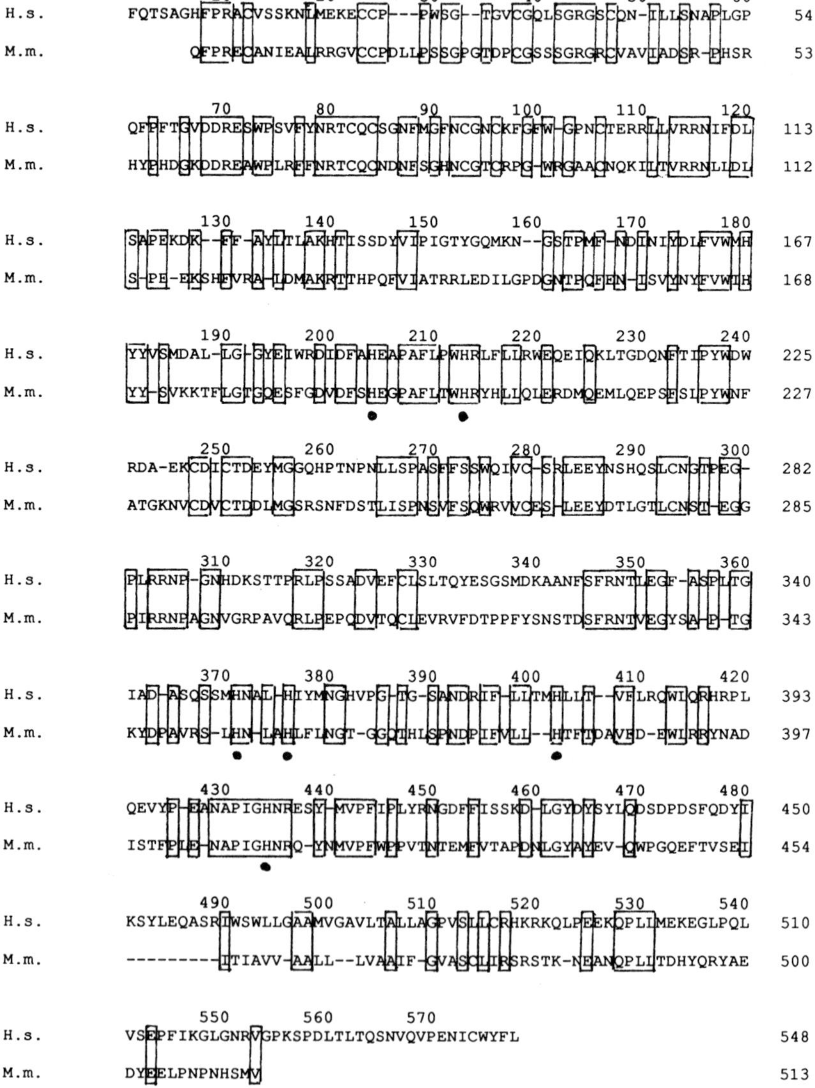

Figure 4.6 Primary sequences of *Homo sapiens* (human; H.s.) and *Mus musculus* (mouse; M.m.; pMT4) polyphenol oxidases. The amino-acid sequences are according to Kwon *et al.* (1987; *H. sapiens*) and Shibahara *et al.* (1986; *M. musculus*). The numbering system is that of *H. sapiens* polyphenol oxidase. Identical sequences are boxed; a dash indicates a space left for alignment purposes.

oxidases and hemocyanins E and D have closely homologous amino-acid sequences in two regions of the proteins (Figure 4.7).

There are two conserved regions, A and B, in all six polyphenol oxidases (Figure 4.7). Region A, consisting of 19 amino-acid residues, has strict conservation of two histidines, a phenylalanine, a tryptophan, an arginine, a leucine and a glutamic acid residue, and conservative homology in other positions. Region B, consisting of up to 42 amino-acid residues, has strict conservation of three histidines, two aspartic acid residues, and one each of asparagine, glycine, serine, proline, phenylalanine and leucine residues and conservative homology in other residues. As already mentioned, this region is very homologous with subunits E

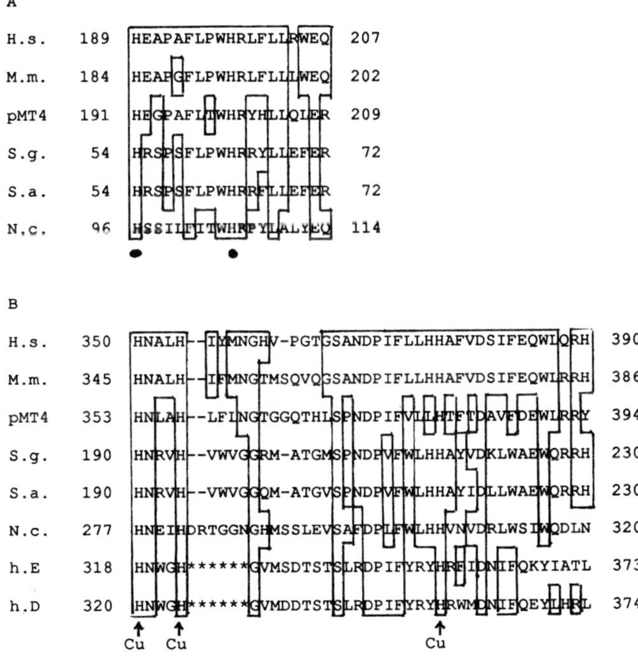

* residues 322–341 and 324–342 of h.E and h.D, respectively, not included here.

Figure 4.7 Homologous regions (A and B) among polyphenol oxidases of several species, the tyrosinase-related protein encoded by pMT4 and two hemocyanins. (A) Amino-acid sequences in conserved region 1. (B) Amino-acid sequences in conserved region 2. The polyphenol oxidase sequence data are from: *H. sapiens* (H.s.), Kwon *et al.* (1987); *M. musculus* (M.m.), Müller *et al.* (1988); pMT4, Shibahara *et al.* (1986); *S. glaucescens* (S.g.) Huber *et al.* (1985); *S. antibioticus* (S.a.), Bernan *et al.* (1985); *N. crassa* (N.c.), Lerch (1982); *Eurypelma californicum* (spider) hemocyanin E (h.E.), Schneider *et al.* (1983); *E. californicum* hemocyanin D (h.D.), Schartau *et al* (1983). The numbers at the beginning and end indicate the amino acid sequence within each polyphenol oxidase. A dash indicates a space left for alignment purposes. Exact amino-acid sequences are boxed. Cu↑ indicates the copper-binding histidines of hemocyanins (Gaykema *et al.*, 1984) (from Müller *et al.*, 1988; with permission.)

(Schneider *et al.*, 1983) and D (Schartau *et al.*, 1983) of hemocyanin of the spider *Eurypelma californicum* and of spring lobster *Panulirus interruptus* (Gaykema *et al.*, 1984).

Histidine residue 306 is known to be important in *N. crassa* polyphenol oxidase from mechanism-based inactivation of the enzyme (Dietler and Lerch, 1982). Homology around Cu(A) is not conserved for all three proteins. Residues 184–197 of *N. crassa* polyphenol oxidase and *E. californicum* hemocyanin have reasonable homology. The two histidines can be aligned (residues 188 and 193 in *N. crassa*; see Figure 4.4). The same histidine residues have been implicated as important in binding copper atoms by photo-oxidation experiments using *N. crassa* polyphenol oxidase (Pfiffer and Lerch, 1981). These two histidine residues are not present in the aligned *S. glaucescens* and *S. antibioticus* polyphenol oxidases; however, there are two histidine residues 96 and 105 that align and have substantial homology around them. Mutant studies in which histidine residues 96 and 278 (see Figure 4.4) have been changed to asparagine show these two histidine residues to be important in copper atom binding (Huber and Lerch, 1988). The third histidine residue bound to Cu(A) may possibly be 296 (see Figure 4.4) based on photo-inactivation experiments (Pfiffer and Lerch, 1981).

There are, however, some substantial differences among the several polyphenol oxidases (Table 4.4). Polyphenol oxidases from *N. crassa* and *S. glaucescens* (and *S. antibioticus*) have 1 and 0 1/2-cystines, while human and mouse have 15 each that are strictly conserved. Polyphenol oxidases from *N. crassa*, *S. glaucescens* and *S. antibioticus* have fewer histidine residues than human and mouse; however, the proteins themselves are considerably smaller. Five histidine residues are strictly conserved in all polyphenol oxidases so far sequenced, with a sixth found at the same location except for pMT4 mouse clone (Müller *et al.*, 1988; see Figure 4.4). The molecular weights range from 30 900 and 46 000 Daltons for *S. glaucescens* and *N. crassa* polyphenol oxidases, respectively, to 58 000 to 62 610

Table 4.4 Some similarities and differences among amino-acid sequences of polyphenol oxidases

Source	1/2-Cys	Number of His Residues	Amino acid number	Molecular weight (Daltons)	Subunits
Neurospora crassa[a]	1	10	407	46 000	1
Streptomyces glaucescens[b]	0	12	273	30 900	1
Human[c]	17	15	548	62 610	1
Mouse[d]	15	16	513	58 000	1
Mouse[e]	15	13	515	58 000	1
Mushroom				128 000	4

[a]Lerch (1983).
[b]Huber *et al.* (1985).
[c]Kwon *et al.* (1987).
[d]Shibahara *et al.* (1986).
[e]Müller *et al.* (1988).

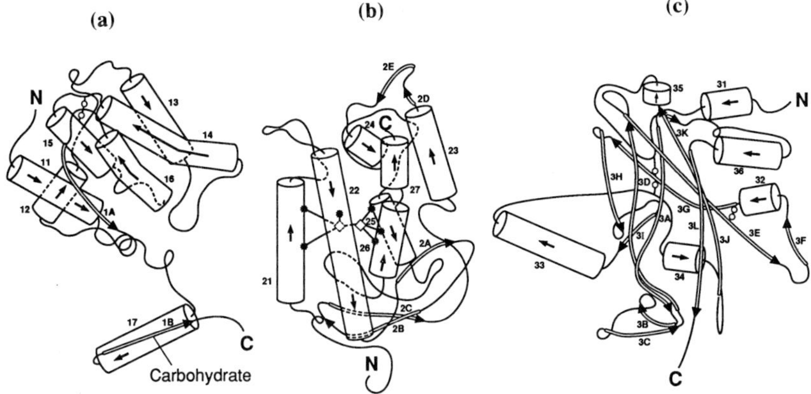

Figure 4.8 The three domains of a subunit of *Panulirus interruptus* (spring lobster) hemocyanin. Domain b contains the two coppers and the six histidine ligands. (From Gaykema *et al.*, 1984 with permission.)

Daltons for mouse and human polyphenol oxidases to 128 000 Daltons for mushroom polyphenol oxidase. All are single polypeptide enzymes except for mushroom polyphenol oxidase, which has four subunits.

The three-dimensional structure of hemocyanins from the spring lobster *Panulirus interruptus* shows each of the two copper atoms, designated as Cu(A) and Cu(B), to be bound by three histidine residues in one of the three domains of the enzyme (see Figure 4.8); Gaykema *et al.*, 1984).

Could the six sequenced polyphenol oxidases have tertiary structures similar to domain b of *P. interruptus* hemocyanin (see Figure 4.8)? Are the histidine residues of the polyphenol oxidases located on adjacent α-helices as in the hemocyanin? Answers await X-ray crystallography studies on one or more of the polyphenol oxidases. It is, however, known that the core of the third hemocyanin domain, consisting of a seven-stranded, antiparallel β-barrel (Gaykema *et al.*, 1984) is identical to that of the β-barrels of Cu, Zn superoxide dismutase (Tainer *et al.*, 1982) and immunoglobulin REI (Epp *et al.*, 1974).

4.4 Control of polyphenol oxidase activity

4.4.1 Breeding, O_2 elimination, heating, pH control

Polyphenol oxidase activity in fruits and vegetables can be controlled in several ways. The levels of polyphenol oxidase in peaches, for example, have been modified by breeding. Some of the best cultivars of Elberta peaches, based on size, color, texture and taste, have high polyphenol oxidase activity, leading to browning even under the most careful handling conditions. While suitable for home gardens, they are not used commercially. As expected from equations

4.1 and 4.2, browning does not occur unless mechanical damage leads to skin breakage and O_2 penetration. Edible films might be useful in this regard to prevent O_2 penetration. Polyphenol oxidase browning in wines, once controlled primarily by sodium bisulfite addition, is now controlled primarily by bottling in an atmosphere of N_2. Polyphenol oxidase activity can be eliminated by heating (blanching); however, this may change the color, texture and taste of the product and obviously cannot be applied to fresh tropical fruits and leafy salads.

Browning can be controlled by adjusting the pH with citric, malic or fumaric acids to pH 4 or below in juices, fruit slices, avocado guacamole, etc., so long as the acidic taste can be tolerated. Lemon juice is frequently used. The pH optima of most plant polyphenol oxidases are near 6.5. At 2.5 pH units below the optimum pH, the enzyme is functioning approximately 1% or less lower than its maximum activity. There may be further reduction in activity below pH 4 due to less-tight binding of copper atoms in the active site of the enzyme, permitting chelators such as citric acid to remove the copper atoms.

4.4.2 Reducing compounds

4.4.2.1 Indirect effects. Ascorbate, isoascorbate, sodium bisulfite and thiol compounds are known to delay the onset of browning. Browning in apple, pear, apricot and peach slices and grapes (golden raisin products) is controlled by placing them in a chamber with burning sulfur where SO_2 and SO_3 are produced. When the gases strike the moist fruit, HSO_3^- and SO_3^{2-} (depending on pH) are formed. In part, browning is prevented by reduction of *o*-benzoquinone formed (see equation 4.2) back to the starting *o*-dihydroxyphenol. In the process, the reducing compound is oxidized. When all the reducing compound is oxidized, no further control of browning occurs by this mechanism. Equation 4.3 shows schematically the principle of this method.

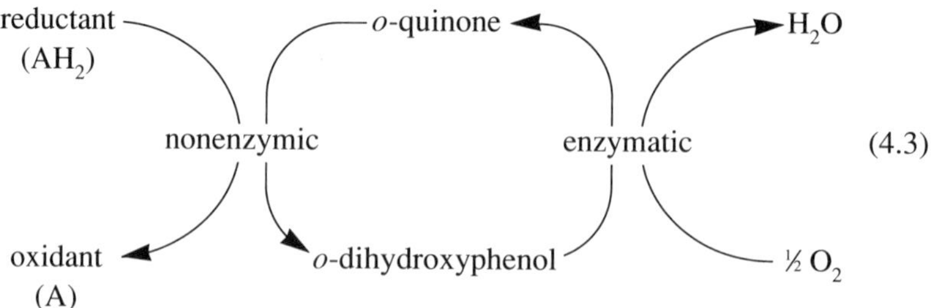

Experimentally, a lag period is observed before any browning occurs. With a fixed concentration of enzyme, the length of the lag period is directly proportional to the concentration of the reductant added (Embs and Markakis, 1965). At one time, the chromomatic method (time until browning) was used to measure polyphenol oxidase activity since, at a fixed concentration of reductant, the lag period was thought to be proportional to enzyme concentration (Dawson and Magee, 1955).

To date, use of reducing compounds, often in combination with each other, is the most effective control method for polyphenol oxidase browning. Sodium bisulfite is most effective followed by thiol compounds and then ascorbate (Golan-Goldhirsh and Whitaker, 1984).

4.4.2.2 Direct effects. From extensive studies with mushroom polyphenol oxidase in the presence and absence of reducing compounds, it appears that ascorbate, bisulfites and thiol compounds have a direct effect on polyphenol oxidase (Golan-Goldhirsh and Whitaker, 1984). This was demonstrated initially by noting that not only was there a lag period before browning proportional to ascorbate concentration but also that the rate of subsequent browning was drastically reduced and that it was dependent on reductant concentration (Figure 4.9). Also, it was found that different substrates affected the lag period and rate of subsequent browning differently, as did the nature of the reducing compound. Experiments were performed in which polyphenol oxidase was incubated with reducing compounds in the absence of substrate and the rates of enzyme inactivation determined. The data in Table 4.5 are a summary of the results for dithiothreitol, sodium bisulfite, glutathione (reduced) and ascorbate in terms of the time required to inactivate half the enzymatic activity. Dithiothreitol was more than 50 times more effective than sodium bisulfite, which was better than glutathione and ascorbate. Subsequently, Golan-Goldhirsh *et al.* (1992) studied the ascorbate reactions in detail.

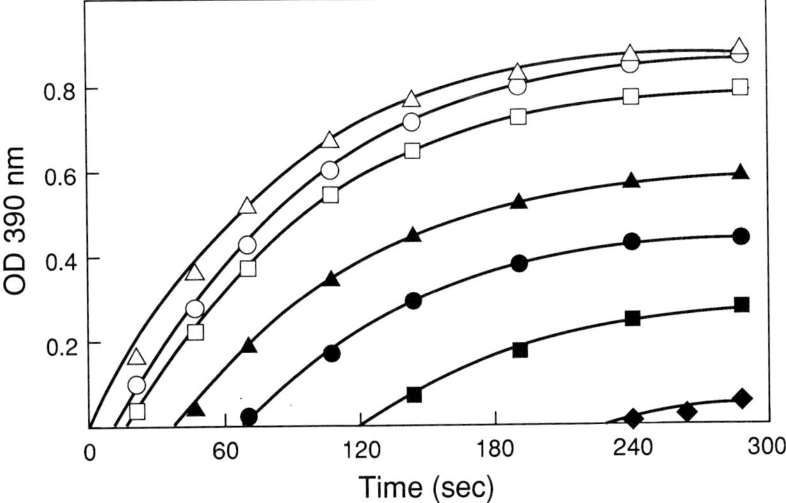

Figure 4.9 Reaction of mushroom polyphenol oxidase (PPO) with pyrocatechol at 23°C in the presence and absence of ascorbic acid. The 1 ml reaction mixture contained 0.1 M sodium phosphate buffer, pH 7.1, 5 mM pyrocatechol, 14 µg/ml mushroom PPO, and the following ascorbic acid concentrations (in mM): (△) 0; (○) 0.1; (□) 0.2; (▲) 0.3; (●) 0.5; (■) 0.7; (◆) 0.8. (From Golan-Goldhirsh and Whitaker, 1984, with permission.)

4.4.2.3 Ascorbate. Freshly prepared ascorbate, at pH 6.5 and 25°C, is relatively ineffective in inactivating mushroom polyphenol oxidase (Golan-Goldhirsh and Whitaker, 1984). With aging, the ascorbate solution becomes more effective and is better then dehydroascorbate, which is more effective than ascorbate. The rate of inactivation was markedly enhanced by addition of Cu^{2+} (as $CuSO_4 \cdot 5H_2O$). At 5 mM ascorbate, the rate increased up to 5 μM Cu^{2+}, but remained constant at 10, 20, 100 and 500 μM Cu^{2+}, showing a saturation effect (Golan-Goldhirsh *et al.*, 1992). Identical results were found for equimolar concentrations of $CuCl_2$, $Cu(NO_3)_2$ and $Cu(C_2H_3O_2)_2$. The rate of inactivation was essentially linear with ascorbate concentration between 1 and 5 mM, showing no saturation effect.

Inactivation occurred only under aerobic conditions (Table 4.6; Golan-Goldhirsh *et al.*, 1992). Gel-permeation chromatography showed that the polyphenol oxidase remained intact but that several new more anionic forms of the enzyme were formed, as detected by polyacrylamide gel electrophoresis. Similar results were found with bovine serum albumin, ovalbumin and Kunitz

Table 4.5 Effect of reductants on polyphenol oxidase (PPO) inactivation[a]

Reductant	$t_{0.5}$ (min)
Dithiothreitol (0.1 mM)	8
Sodium bisulfite (5 mM)	28
Glutathione (5 mM)	106
Ascorbate (5 mM)	130

[a]Adapted from Golan-Goldhirsh and Whitaker (1984). Activity loss determined polarographically; 6.3 μM mushroom polyphenol oxidase, in 0.11 M phosphate buffer, pH 6.5, 25°C.

Table 4.6 Requirement of O_2 for polyphenol oxidase (PPO) inactivation by ascorbate and copper

Condition[a]	Incubation time (min)	Activity remaining (%)
Aerobic		
PPO	1	100
PPO, ascorbate/Cu^{2+},[b]	1	100
	60	28
Anaerobic[b,c]		
PPO, ascorbate/Cu^{2+}	1	100
	60	100
	114[d]	30

[a]Adapted from Golan-Goldhirsh *et al.* (1992). 7.1 μM Mushroom polyphenol oxidase in 0.11 M phosphate buffer, pH 6.5 and 25°C.
[b]5 mM Ascorbate, 0.8 mM $CuSO_4 \cdot 5H_2O$.
[c]All solutions were deaerated thoroughly under vacuum before mixing and the system was continually flushed with N_2.
[d]After 60 min under N_2, aerobic conditions restored.

soybean trypsin inhibitor (Golan-Goldhirsh *et al.*, 1992), indicating that the reactions are not unique to polyphenol oxidase.

Amino-acid analyses of polyphenol oxidase showed that there was a major decrease in the histidine content, a smaller decrease in the methionine content and an increase in aspartic acid, glutamic acid, glycine and alanine (Table 4.7). Analysis using urease showed that urea was formed in nearly stoichiometric amounts with aspartic acid. Hydrogen peroxide was also formed, as shown by peroxidase/guaiacol assays. Similar results were found following ascorbate/Cu^{2+} treatment of bovine serum albumin, ovalbumin and Kunitz soybean trypsin inhibitor. For example, 15 of the 17 histidine residues of bovine serum albumin were lost, with an increase from 51 to 61 aspartic acid/asparagine residues, 74 to 81 glutamic acid/glutamine residues, 15 to 19 glycine residues and 39 to 46 alanine residues (Golan-Goldhirsh *et al.*, 1992).

How can one increase the aspartic acid, glutamic acid, glycine and alanine content of a protein, while at the same time accounting for major decreases in the histidine content? Previously, Imanaga (1955) showed that treatment of histidine derivatives with ascorbate and Cu^{2+} gave a number of products, the major one being aspartic acid. Photo-oxidation of histidine yields similar products. Oxidative scission of the imidazole side-chain of histidine gives aspartic acid, glutamic acid, alanine or glycine, plus other products, when scission occurs at a–e, b–e, c–e, or d–e, respectively (Figure 4.10), following hydrolysis of the protein. Aspartic acid is the major product formed from ascorbate and Cu^{2+} treatment of polyphenol oxidase at 25°C and pH 6.5 (see Table 4.7), although there are significant changes in glutamic acid/glutamine, glycine and alanine. None of the other amino acids showed changes.

The reactive components are the imidazole side-chain of histidine, ascorbate (probably as the semi-dehydroascorbate free radical), copper (probably as Cu^{2+} locally) and O_2. Golan-Goldhirsh *et al.* (1992) proposed that one role of Cu^{2+} is

Table 4.7 Amino-acid composition of polyphenol oxidase (PPO) incubated with ascorbate/Cu^{2+},[a]

	PPO	
Amino acid	Control[b]	Treated[c]
His	19.5	5.1
Met	7.0	4.2
Asx	127	157
Glx	125	145
Gly	92.4	108
Ala	97.0	111

[a]Adapted from Golan-Goldhirsh *et al.* (1992). Moles amino acid per mol. protein; 130 000 Daltons.
[b]Reaction conditions: 19.0 µM protein, 0.1 M sodium phosphate buffer, pH 6.5, 25°C for 22 h in a gyrotary water bath shaker.
[c]As in footnote *b* except contained 5 mM ascorbate and 0.8 mM $CuSO_4$.

to bind to the imidazole ring, O_2 and semi-dehydroascorbate, so that the reaction takes place specifically at this site. Ascorbate or Cu^{2+}, separately, are relatively ineffective in oxidizing the histidine residues in polyphenol oxidase (Table 4.8). Evidence for the proposed mechanism is as follows:

1. Ascorbate or dehydroascorbate is less effective than some intermediate formed during incubation of ascorbate in air.
2. Ascorbate/Cu^{2+} is most effective.
3. Cu^{2+} alone is rather ineffective.
4. O_2 is required.
5. Histidine specifically is lost.
6. Aspartic acid, glutamic acid, alanine and glycine formation is indicative of the free-radical mechanism.
7. Loss of histidine does not involve protein chain cleavage.

It has been demonstrated by electron spin resonance (ESR) that free radicals are formed when ascorbate and Cu^{2+} are mixed in the presence of O_2 (Figure 4.11; Uchida and Kawakishi, 1986). However, the authors have been unsuccessful using trapping experiments to determine whether the semi-dehydroascorbate free radical, O_2^-, HO·, HO$_2$· or H_2O_2 is the effective oxidant, because of the local production of the oxidant. Hydrogen peroxide added at 100 times the concentration produced in the reaction had no effect on enzyme activity. Other workers have suggested that HO· is the major oxidant, based on its oxidant strength but have not offered

Table 4.8 Histidine content of mushroom polyphenol oxidase (PPO) incubated with ascorbate, copper or ascorbate/copper[a]

Conditions	Number of His residues[b]
Control	18.6
Ascorbate (5 mM)	18.6
Cu^2 (0.8 mM)	18.6
Ascorbate/Cu^{2+} (5 mM/0.8 mM)	4.6

[a]Adapted from Golan-Goldhirsh *et al.* (1992). 19 µM protein, 0.1 M sodium phosphate buffer, pH 6.5, 25°C, 22 h.
[b]Per molecular weight of 130 000 Daltons.

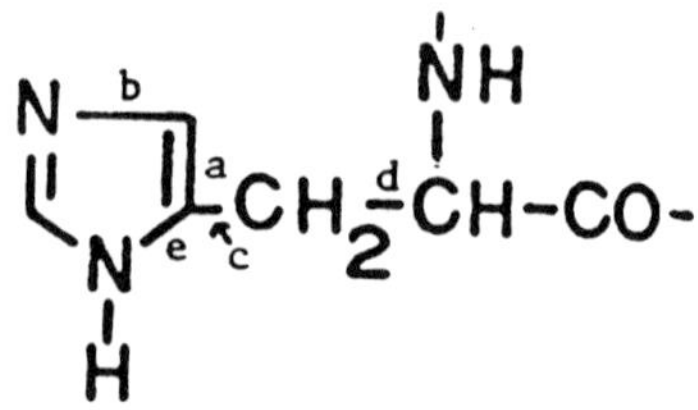

Figure 4.10 Proposed scission of histidine residue by ascorbic acid/Cu^{2+} to give aspartic acid, (a–e), glutamic acid (b–e), alanine (c), glycine (d) and urea (a–e).

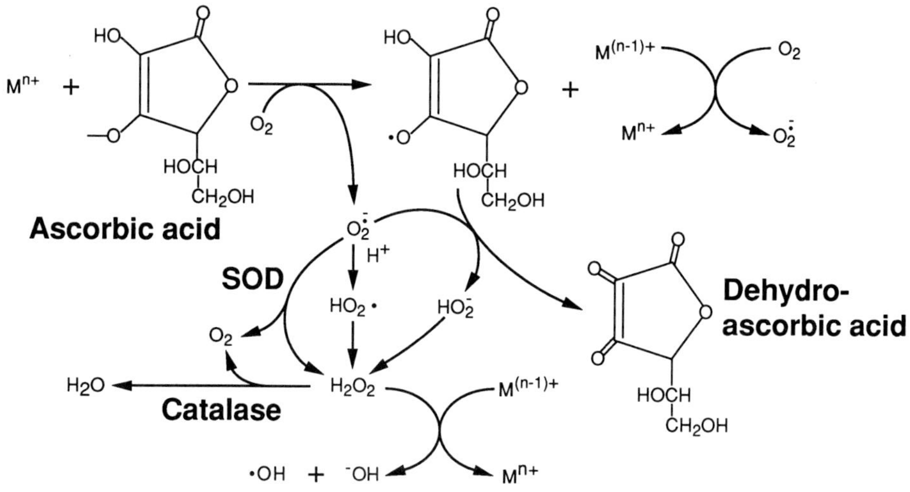

Figure 4.11 Formation of free radicals during reaction of ascorbate, metal ions and O_2 (from Uchida and Kawakishi, 1986, with permission).

experimental verification (Dietler and Lerch, 1982). Detailed ESR studies, using the latest ESR instrumentation interfaced with computers, are clearly needed.

Copper ions are released from the polyphenol oxidase during inactivation (Mason *et al.*, 1961; Dietler and Lerch, 1982) presumably due to formation of a free-radical intermediate, the semi-benzoquinone radical. Photo-oxidation in the presence of citrate causes inactivation of mushroom polyphenol oxidase, with release of copper ions (Fry and Strothkamp, 1983).

4.4.2.4 Thiol compounds. Thiol compounds, such as dithiothreitol, dithioery-thritol and cysteine, are effective inactivators of polyphenol oxidase (see Table 4.5; Golan-Goldhirsh *et al.*, 1992). Direct inactivation of mushroom polyphenol oxidase could be the result of: (i) reduction of one or more essential disulfide bonds in the enzyme; (ii) reduction of ECu(II) to ECu(I) – the Cu(I) then dissociates from the enzyme quite readily or (iii) the removal of Cu(II) from the enzyme without prior reduction to Cu(I) due to RSCu(II)SR formation (Lerch, 1987). The authors incubated mushroom polyphenol oxidase (39.6 μM) with dithioerythritol (2 mM) in pH 6.5 (0.1 M) sodium phosphate buffer, at 23°C, in the presence of 1 mM EDTA. After 155 min, the activity decreased to 20% of the initial value. The treated enzyme was rapidly purified by Sephadex gel-permeation chromatography equilibrated with N_2-purged 0.1 M phosphate buffer, pH 6.5, containing EDTA. There was no increase in activity. No increase in the number of –SH groups was found, as determined by the 5,5'-dithiobis-(2-nitrobenzoic acid) method of Ellman (1959). The Cu^{2+} content of the treated enzyme was decreased by approximately 80%, as determined by atomic absorbance spectroscopy; however, incubation of the purified treated enzyme with Cu^{2+} restored most of its activity. Therefore, it is concluded that thiol compounds inactivate polyphenol oxidase by reducing Cu(II) to Cu(I), which then readily dissociates from the enzyme (Martell and Smith, 1974; Smith and Martell, 1975). Half-met-polyphenol oxidase ((Cu(I)–Cu(II)) can be formed with *N. crassa* polyphenol oxidase (Wilcox *et al.*, 1985), and presumably *Agaricus bisporus* (mushroom) polyphenol oxidase (Schoot Uiterkamp and Mason, 1973) by treating the enzyme with $NaNO_2$ and ascorbate at pH 6.3. This is a further indication that the binuclear copper atoms of the active site are accessible to reducing compounds.

4.4.2.5 Sodium bisulfite.   Sodium bisulfite is an effective inhibitor of polyphenol oxidase (Embs and Markakis, 1965; Golan-Goldhirsh and Whitaker, 1984). It has been widely used during preparation of fresh salads and wines and is still used during drying of golden raisins and apricot, peach and apple slices. Unfortunately, some people can detect bisulfite at <100 ppm (0.1 mM) and persons with respiratory difficulties may be dangerously affected, probably by SO_2. Sodium bisulfite decreases the rate of polyphenol oxidase browning by reducing the *o*-benzoquinone back to the *o*-dihydroxyphenol (Embs and Markakis, 1965). In addition, sodium bisulfite has a direct effect on polyphenol oxidase, causing rapid inactivation (Golan-Goldhirsh and Whitaker, 1985; see Table 4.5). Possible explanations for the direct inactivation of polyphenol oxidase include (i) sulfitolysis of one or more disulfide bonds that are essential for activity of the enzyme; (ii) reduction of ECu(II) to ECu(I), permitting the Cu(I) to dissociate readily from the enzyme; and (iii) the removal of Cu(II) from the enzyme by formation of $Cu(HSO_3)_2$. The authors have shown that sulfitolysis of the disulfide bonds does not occur with the native mushroom enzyme. Only when the enzyme is previously denatured does sulfitolysis of disulfide bonds occur. It has also been shown that $Cu(HSO_3)_2$ is essentially completely ionized at pH 6.5 and, therefore, could not compete with the enzyme for Cu(II). In model systems sodium bisulfite rapidly reduces Cu(II) to Cu(I). Furthermore incubation of mushroom polyphenol oxidase with sodium bisulfite until the enzymatic activity is reduced to 10–20% the original value results in approximately 80% loss of copper ions from the enzyme. Some of the isozymes of mushroom polyphenol oxidase were more sensitive to sodium bisulfite than others. Therefore, it is concluded that sodium bisulfite inactivates mushroom polyphenol oxidase by reducing enzyme bound Cu(II) to Cu(I), which then dissociates from the enzyme.

4.4.2.6 k_{cat} inactivation.   Polyphenol oxidase is rapidly inactivated during reaction with substrate (Figure 4.12). Addition of more *o*-dihydroxyphenol or O_2 (when the initial reaction includes bubbling O_2 through the solution before adding enzyme) does not restore activity. Only addition of a new aliquot of enzyme results in additional product formation. (The decrease in absorbance on the right side of the curve in Figure 4.12 simply indicates that the λ_{max} shifts as

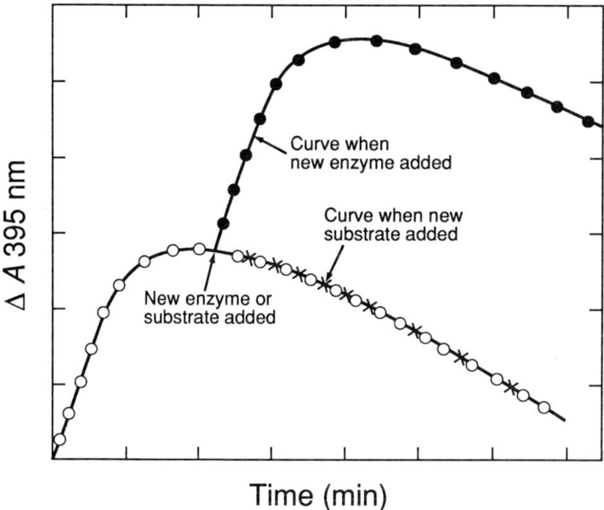

Figure 4.12 Rate of oxidation of catechol by mushroom polyphenol oxidase (○). At time indicated by arrow, new substrate (×) or new enzyme (●) was added to the reaction in a new experiment.

melanin is formed.) Associated with the loss of enzyme activity is an increase in free copper ions (released from the inactive enzyme) and a decrease in histidine (Dietler and Lerch, 1982). There is direct evidence that a free-radical intermediate, o-semibenzoquinone, is formed in the reaction (Mason et al., 1961) and that the free radical, while still bound in the active site, causes scission of the imidazole side-chain of histidine, analogous to the ascorbate/Cu^{2+} reaction described above. The reactions involved are shown in equation 4.4.

$$2 \text{ (catechol)} + 1/2\ O_2 \xrightarrow[H_2O]{PPP} 2 \text{ (semiquinone)} \xrightarrow[H_2O]{PPO;\ 1/2\ O_2} 2 \text{ (quinone)} \tag{4.4}$$

It is estimated that about one enzyme molecule is inactivated for each 5 000 conversions of substrate to product (Golan-Goldhirsh and Whitaker, 1985). This is not an effective k_{cat} inactivation of the enzyme in relation to the rate of browning and inactivation is not due to melanin formation and subsequent binding to the enzyme during the early part of the reaction (see Figure 4.12) during the 90–120 s reaction time.

The authors reasoned that if the free radical intermediate in the active site of intermediate much faster than its disappearance to o-benzoquinone), the efficiency of inactivation of the enzyme could be increased. Perhaps this could be achieved by changing the nature of the group in the 4-position of the catechol (see equation 4.4). This idea was tested using catechol, 4–methylcatechol and o-dihydroxyphenylalanine (DOPA) at concentrations $10 \times K_m$ for each substrate. The reactions were run at 25°C and pH 6.5 in 0.1 M sodium phosphate buffer. Identical rates of inactivation of polyphenol oxidase for the three substrates were observed. Therefore, the rate-limiting step appears to be formation of the free radical.

4.4.2.7 Removal of phenols. It should be obvious that removal of phenol substrates from foods would prevent browning. This is simple enough in principle but difficult in practice; it is easier to exclude the O_2. With juices, including wines, several exclusion methods have potential. The juices, if free of suspended materials (largely pectin), can be passed through columns of Sephadex G-25, which bind phenols or columns of immobilized cyclodextrins, occluding some phenols. The juices can also be treated batchwise with insoluble Sephadex G-25, cyclodextrins, polyethylene glycol or polyvinyl pyrrolidone, which bind phenols, then filtered. Sapers and Hicks (1989) studied the effect of cyclodextrins in preventing browning and have a patent pending on the use of cyclodextrins for this purpose.

The authors tested the ability of cyclodextrins to bind some phenols that serve as substrates for polyphenol oxidase, thereby effectively reducing the substrate concentration available to the enzyme according to equation 4.5 (see Table 4.9).

The K_d values for the cyclodextrins (CD) were calculated from plots of $1/v_o$ versus $1/[S]_o$ in the presence and absence of added CD. The CD, at concentrations in the range of K_d, behaved as competitors with the enzyme for binding to substrate, when the substrate was in the range of 0.05–5 K_m. At $[S]_o \gg K_m$ no

Table 4.9 Dissociation constants (K_d) for polyphenol oxidase substrates to cyclodextrins[a]

Substrate	K_d (mM)[b]	
	β-cyclodextrin	α-cyclodextrin
Chlorogenic acid	2.67	2.0
Caffeic acid	6.24	9.78
Dopamine	53.3	[c]
Catechol	[c]	[c]
DL-dihydroxyphenylalanine	[c]	[c]

[a]A. van der Schaaf and J.R. Whitaker (1994), unpublished data. The reactions contained 0.22 μM mushroom polyphenol oxidase, substrate ranging from 0.05–5 K_m, 0, 5 or 10 mM cyclodextrin in 0.1 M phosphate buffer, pH 6.5, at 22°C.
[b]K_d was calculated from plots of $1/v_o$ versus $1/[S]_o$ in the presence and absence of a fixed concentration of cyclodextrin. v_o is the initial velocity of the reaction recorded continuously on a recording spectrophotometer at λ_{max} for the o-benzoquinone formed.
[c]No decrease of v_o was found in the presence of cyclodextrin indicating that K_d was infinite.

$$E + S \underset{\xleftarrow{\hspace{1em}}}{\overset{K_s}{\xrightarrow{\hspace{1em}}}} E \cdot S \xrightarrow{\quad k_2 \quad} E + P$$

$$K_d \downharpoonright\downharpoonright \; CD$$

$$S \cdot CD$$

(4.5)

effect of CD on v_0 was observed. As noted, K_d values are in the 2–10 mM range for chlorogenic acid and caffeic acid for both α- and β-cyclodextrins, and 50 mM range for dopamine and β-cyclodextrin. Dopamine did not appear to bind to α-cyclodextrin and catechol and DL-dihydroxyphenylalanine did not appear to bind to either α- or β-cyclodextrin, as shown by the lack of effect on v_0.

Unexpectedly, increased ionic strength (with NaCl) of the solution increased K_d for chlorogenic acid binding to β-cyclodextrin ([NaCl], K_d: 0 M, 2.67 mM; 0.10 M, 10.0 mM; 0.20 M, 11.6 mM; 0.33 M, 26.0 mM) but had little effect on K_d for caffeic acid binding to β-cyclodextrin ([NaCl], K_d: 0 M, 6.2 mM; 0.20 M, 5.62 mM; 0.33 M, 5.09 mM). As expected, ethanol added to the reaction increased K_d. The results were: chlorogenic acid and β-cyclodextrin, 2.67 mM, 8.13 mM, and 8.08 mM for 0%, 5% and 7% ethanol, respectively; caffeic acid and β-cyclodextrin, 6.24 mM, 14.9 mM and 27.4 mM for 0%, 5% and 7% ethanol, respectively.

Binding of polyphenol oxidase substrates to polyvinylpyrrolidone (PVP) was also measured by the soluble PVP effect on v_0. The same experimental conditions were used as for the α- and β-cyclodextrins studies. PVP was a competitive inhibitor with enzyme for binding to substrate. The calculated K_d values, based on the effect of added PVP on v_0, were 129 mM for chlorogenic acid and PVP, and 243 mM for caffeic acid and PVP. The concentration of PVP was expressed on an equivalent number of vinylpyrrolidone units in PVP (259 units/10 000 Daltons). Therefore, PVP is a much less effective binder to chlorogenic acid and caffeic acid than is β-cyclodextrin. More effective adsorbants of phenols are needed, if they are to be used to prevent polyphenol oxidase-catalyzed browning.

Browning can be prevented by methylation of the hydroxy group of monophenols with o-methyltransferase (Finkle and Nelson, 1963; Finkle, 1964). However, this is not yet commercially feasible, or economically affordable at the present time.

4.4.2.8 Competitive inhibitors. Benzoic acid and other benzene derivatives are good competitive inhibitors of polyphenol oxidase (Lerch, 1987). Care must be taken to select competitive inhibitors that do not react with o-benzoquinones via Michael addition, leading to highly colored products (Wong and Whitaker, 1971b). Values of K_i (dissociation constant for E+I $\Leftrightarrow$ EI) as low as 1 μM for benzoic acid have been reported; however, Rivas and Whitaker (1973) reported K_i values of about 10 μM for peach polyphenol oxidases and the authors of this chapter found K_i values of about 10 μM for mushroom polyphenol oxidase, using catechol or 4-methylcatechol as substrate.

Recently, it has been shown that diamine derivatives of coumarin and 4-hexylresorcinol are effective inhibitors of black-spot formation in shrimp, with K_i values of approximately 0.1–10 μM (McEvily *et al.*, 1991). The authors of this chapter found K_i to be 0.1 μM for 4-hexylresorcinol and mushroom polyphenol oxidase acting on *o*-dihydroxyphenylalanine (DOPA) at pH 6.5 and 25°C. The authors are now testing 4-hexylresorcinol for the control of polyphenol oxidase in grape juice. The pH and temperature dependence of K_i for these inhibitors need to be determined, as well as their stability during long-term storage of wines and juices.

These more-effective reversible inhibitors that are found in plant materials (diamine derivatives of coumarin) and which are used in over-the-counter pharmaceuticals (4-hexylresorcinol in Sucrets) offer good opportunities for control of polyphenol oxidase in juices, wines and other products.

Acknowledgment

The authors thank Bilateral Agricultural Research Development (BARD) for support of much of this research in Project No. I-1074-86.

References

Bernan, V., Filpula, D., Herber, W., Bibb, M. and Katz, E. (1985) The nucleotide sequence of the tyrosinase gene from *Streptomyces antibioticus* and characterization of the gene product. *Gene* **37**:101–10.

Bouchilloux, S., McMahill, P. and Mason, H.S. (1963) The multiple forms of mushroom tyrosinase. Purification and molecular properties of the enzymes. *J. Biol. Chem.* **238**:1699–1707.

Cleland, W.W. (1963a) The kinetics of enzyme-catalyzed reactions with two or more substrates or products. I. Nomenclature and rate equations. *Biochim. Biophys. Acta* **67**:104–37.

Cleland, W.W. (1963b) The kinetics of enzyme-catalyzed reactions with two or more substrates or products. II. Inhibition: nomenclature and theory. *Biochim. Biophys. Acta* **67**:173–87.

Cleland, W.W. (1963c) The kinetics of enzyme-catalyzed reactions with two or more substrates or products. III. Prediction of initial velocity and inhibition patterns by inspection. *Biochim. Biophys. Acta* **67**:188–96.

Dawson, C.R. and Magee, R.J. (1955) Plant tyrosinase (polyphenol oxidase). *Methods Enzymol.* **2**:817–27.

Dietler, C. and Lerch, K. (1982) Reaction inactivation of tyrosinase, in *Oxidases and Related Redox Systems* (eds T.E. King, H.S. Mason and M. Morrison), Proceedings of the Third International Symposium, 1979. Pergamon, Oxford, pp. 305–17.

Duckworth, H.W. and Coleman, J.E. (1970) Physicochemical and kinetic properties of mushroom tyrosinase. *J. Biol. Chem.* **245**:1613–25.

Ellman, G.L. (1959) Tissue sulfhydryl groups. *Arch. Biochem. Biophys.* **82**:70–7.

Embs, R.J. and Markakis, P. (1965) The mechanism of sulfite inhibition of browning caused by polyphenol oxidase. *J. Food Sci.* **30**:753–58.

Enzyme Nomenclature (1979) *Recommendations (1978) of the Nomenclature Committee of the International Union of Biochemistry.* Academic Press, New York.

Epp, O., Colman, P., Fehlhammer, H., Bode, W., Schiffer, M. and Huber, R. (1974) Crystal and molecular structure of a dimer composed of the variable portions of the Bence-Jones protein REI. *Eur. J. Biochem.* **45**:513–24.

Eur. J. Biochem. **45**:513–24.

Finkle, B.J. (1964) Treatment of plant tissue to prevent browning. US Patent No. 3/126/287, 24 March.

Finkle, B.J. and Nelson, R.F. (1963) Enzyme reactions with phenolic compounds: Effect of 0-methyltransferase on a natural substrate of fruit polyphenoloxidase. *Nature* **197**:902–3.

Fling, M., Horowitz, N.H. and Heinemann, S.F. (1963) The isolation and properties of crystalline tyrosinase from *Neurospora*. *J. Biol. Chem.* **238**:2045–53.

Fry, D.C. and Strothkamp, K.G. (1983) Photoinactivation of *Agaricus bisporus* tyrosinase: Modification of the binuclear copper site. *Biochemistry* **22**:4949–53.

Gaykema, W.P.J., Hol, W.G.J., Vereijken, N.M., Soeter, M.N., Bak, H.J., and Beintema, J.J. (1984) 3.2 Å Structure of the copper-containing, oxygen-carrying protein *Panulirus interruptus* haemocyanin. *Nature* **309**:23–29.

Golan-Goldhirsh, A. and Whitaker, J.R. (1984) Effect of ascorbic acid, sodium bisulfite, and thiol compounds on mushroom polyphenol oxidase. *J. Agric. Food Chem.* **32**:1003–9.

Golan-Goldhirsh, A. and Whitaker, J.R. (1985) k_{cat} inactivation of mushroom polyphenol oxidase. *J. Mol. Catal.* **32**:141–47.

Golan-Goldhirsh, A., Osuga, D.T., Chen, A.O. and Whitaker, J.R. (1992) Effect of ascorbic acid and copper on proteins and other polymers, in *The Bioorganic Chemistry of Enzymatic Catalysis: An Homage to Myron L. Bender*, (eds V.T. D'Souza and J. Feder), CRC Press, Boca Raton, pp. 61–76.

Himmelwright, R.S., Eickman, N.C., LuBien, C.D., Lerch, K. and Solomon, E.I. (1980) Chemical and spectroscopic studies of the binuclear copper active site of *Neurospora* tyrosinase: comparison to hemocyanin. *J. Amer. Chem. Soc.* **102**:7339–44.

Hochstein, P. and Cohen, G. (1963) The cytotoxicity of melanin precursors. *Ann. NY Acad. Sci.* **160**:876–86.

Huber, M. and Lerch, K. (1988) Identification of two histidines as copper ligands in *Streptomyces glaucescens* tyrosinase. *Biochem.* **27**:5610–15.

Huber, M. Hinterman, G. and Lerch, K. (1985) Primary Structure of tyrosinase from *Streptomyces glaucescens. Biochem* **24**:6038–41.

Hunt, M.D., Eannetta, N.T., Yu, H., Newman, S.M. and Steffens, J.C. (1993) cDNA cloning and expression of potato polyphenol oxidase. *Plant Mol. Biol.* **21**:59–68.

Imanaga, Y. (1955) Autoxidation of L-ascorbic acid and imidazole nucleus. II. The decomposition products of imidazole derivatives present in the autoxidation mixture. *J. Biochem.* **42**:669–76.

Jolley, Jr, R.L., Nelson, R.M. and Robb, D.A. (1969) The multiple forms of mushroom tyrosinase. *J. Biol. Chem.* **244**:3251–57.

Jolley, Jr, R.L., Evans, L.H., Makino, N. and Mason, H.S. (1974) Oxytyrosinase. *J. Biol. Chem.* **249**:335–45.

Kwon, B.S., Haq, A.K., Pomerantz, S.H. and Halaban, R. (1987) Isolation and sequence of a cDNA clone for human tyrosinase that maps at the mouse c-albino locus. *Proc. Natl Acad. Sci. USA* **84**:7473–77.

Lerch, K. (1978) Amino-acid sequence of tyrosinase from *Neurospora crassa*. *Proc. Natl Acad. Sci. USA* **75**:3635–39.

Lerch, K. (1981) Copper monooxygenases: tyrosinase and dopamine β-monooxygenase. *Metal Ions Biol. Syst.* **13**:143–86.

Lerch, K. (1982) Primary structure of tyrosinase from *Neurospora crassa*. II. Complete amino-acid sequence and chemical structure of a tripeptide containing an unusual thioether. *J. Biol. Chem.* **257**:6414–19.

Lerch, K. (1983) *Neurospora* tyrosinase: structural, spectroscopic and catalytic properties. *Mol. Cell. Biochem.* **52**:125–38.

Lerch, K. (1987) Monophenol monoxygenase from *Neurospora crassa*. *Methods Enzymol.* **142**:165–69.

Lerch, K. and Ettlinger, L. (1972) Purification and characterization of a tyrosinase from *Streptomyces glaucescens*. *Eur. J. Biochem.* **31**:427–37.

Martell, A.E. and Smith, R.M. (1974). *Critical Stability Constants*, Vol. 1, Plenum Press, New York.

Mason, H.S. (1965). Oxidases. *Annu. Rev. Biochem.* **34**:595–634.

Mason, H.S., Spencer, E. and Yamazaki, I. (1961) Identification by electron-spin resonance spectroscopy of the primary product of tyrosinase. *Biochem. Biophys. Res. Commun.* **4**:236–38.

McEvily, A.J., Iyengar, R. and Gross, A. (1991). *Inhibition of Polyphenol Oxidase by Phenolic Compounds*. The Fourth Chemical Congress of North America, 25–30 August 1991, Abstract

Müller, G., Ruppert, S., Schmid, E. and Schütz, G. (1988) Functional analysis of alternatively spliced tyrosinase gene transcripts. *EMBO J.* **7**:2723–30.

Pfiffer, E. and Lerch, K. (1981) Histidine at the active site of *Neurospora* tyrosinase. *Biochemistry* **20**:6029–35.

Pomerantz, S.H. (1963) Separation, purification and properties of two tyrosinases from hamster melanoma. *J. Biol. Chem.* **238**:2351–57.

Rivas, N.J. and Whitaker, J.R. (1973) Purification and some properties of two polyphenol oxidases from Bartlett pears. *Plant Physiol.* **52**:501–7.

Sapers, G.M. and Hicks, K.B. (1989) Inhibition of enzymatic browning in fruits and vegetables, in *Quality Factors of Fruits and Vegetables*, (ed. J.J. Jen.) *ACS Symposium Series* 405, pp. 29–43.

Schartau, W., Eyerle, F., Reisinger, P., Geisert, H., Storz, H. and Linzen, B. (1983) Hemocyanin in spiders. XIX. Complete amino-acid sequence of subunit d from *Eurypelma californicum* hemocyanin, and comparison to chain e. *Hoppe-Seyler's Z. Physiol. Chem.* **364**:1383–1409.

Schneider, H.J., Drexel, R., Feldmaier, G. and Linzen. B. (1983) Hemocyanins in spiders, XVIII. Complete amino-acid sequence of subunit e from *Eurypelma californicum* hemocyanin. *Hoppe-Seyler's Z. Physiol. Chem.* **364**:1357–81.

Schoot Uiterkamp, A.S.M. and Mason, H.S. (1973) Magnetic dipole–dipole coupled Cu(II) pairs in nitric-oxide-treated tyrosinase: A structural relationship between the active sites of tyrosinase and hemocyanin. *Proc. Natl Acad. Sci. USA* **70**:993–96.

Shahar, T., Henning, N., Gutfinger, T., Hareven, D. and Lifschitz, E. (1992) The tomato 66.3-kD polyphenoloxidase gene: molecular identification and developmental expression. *The Plant Cell*, **4**:135–147.

Shibahara, S., Tomita, Y., Sakakura, T., Nager, C., Chaudhuri, B. and Müller, R. (1986) Cloning and expression of cDNA encoding mouse tyrosinase. *Nucl. Acids Res.* **14**:2413–27.

Solomon, E.I. (1981) Binuclear copper active site. Hemocyanin, tyrosinase, and type 3 copper oxidases, in *Copper Proteins*, (ed. T.G. Spiro), Wiley, New York, pp. 41–108.

Smith, R.M. and Martell, A.E. (1975) *Critical Stability Constants*, Vol. 3, Plenum Press, New York.

Sussman, A.S. (1961). A comparison of the properties of two forms of tyrosinase from *Neurospora crassa*. *Arch. Biochem. Biophys.* **95**:407–15.

Szent-Györgyi, A. and Vietorisz, K. (1931) Function and significance of polyphenol oxidase from potatoes. *Biochem. S.* **233**:236–39.

Tainer, J.A., Getzoff, E.D., Beem, K.M., Richardson, J.S. and Richardson, D.C. (1982) Determination and analysis of the 2Å structure of copper, zinc superoxide dismutase. *J. Mol. Biol.* **160**:181–217.

Uchida, K. and Kawakishi, S. (1986) Oxidative degradation of β-cyclodextrin by an ascorbic acid-copper ion system. *Agric. Biol. Chem.* 50:367–73.

Wilcox, D.E., Porras, A.G., Hwang, Y.T., Lerch, K., Winkler, M.E. and Solomon, E.I. (1985) Substrate analogue binding to the coupled dinuclear copper active site in tyrosinase. *J. Amer. Chem. Soc.* **107**:4015–27.

Winkler, M.E., Lerch, K. and Solomon, E.I. (1981) Competitive inhibitor binding to the binuclear copper active site in tyrosinase. *J. Amer. Chem. Soc.* **103**:7001–2.

Witkop, Jr, C.J. (1984) Inherited disorders of pigmentation, in *Genodermatoses: Clinics in Dermatology*, (ed. R.M. Goodman), J.B. Lippincot. Philadelphia, Vol. 2, pp. 70–134.

Wong, T.C., Luh, B.S. and Whitaker, J.R. (1971a) Isolation and characterization of polyphenol oxidases of clingstone peach. *Plant Physiol.* **48**:19–23.

Wong, T.C., Luh, B.S. and Whitaker, J.R. (1971b) Effect of phloroglucinol and resorcinol on the clingstone peach polyphenol oxidase-catalyzed oxidation of 4-methylcatechol. *Plant Physiol.* **48**:24–30.

Yasunobu, K.T. (1959) Mode of action of tyrosinase, in *Pigment Cell Biology*, (ed. M. Gordon), Academic Press, New York, pp. 583–608.

5 Naturally occurring α-amylase inhibitors: Structure/function relationships

M.F. Ho, X. Yin, F.F. Filho, F. Lajolo and J.R. Whitaker

Abstract

Perhaps because α-amylase is ubiquitous in all living systems, several natural inhibitors against this enzyme are found in biological materials. These inhibitors include the microbial nitrogen-containing carbohydrates with an oligobioamine unit, the microbial polypeptides such as Paim and Haim, and the larger protein inhibitors, found in cereals, legumes and some other higher plants. The protein inhibitors were discovered as early as 1933; however, much of the research has been performed since the mid 1970s, with important medical, nutritional and insect-control implications. The structures of several of the microbial N-containing carbohydrates and microbial polypeptides are known, and amino-acid sequences are also known for some of the higher plant protein inhibitors. Enzyme specificity studies with several α-amylases have been carried out, some detailed kinetic studies are available and a few chemical modification studies have been performed. Nevertheless, very little is known about why and how the compounds inhibit α-amylases, primarily those from animals and insects.

5.1 Introduction

Naturally occurring α-amylase inhibitors were reported in wheat in 1933 (Chrzaszcz and Janicki, 1933, 1934), rediscovered in 1943 (Kneen and Sandstedt, 1943, 1946) and again in 1973 (Saunders and Lang, 1973; Silano *et al.*, 1973). Since then, many publications on α-amylase inhibitors in wheat have appeared. Indeed, α-amylase inhibitors were reported in beans in 1945 (Bowman, 1945), and investigated again in 1968 and 1971 (Hernandez and Jaffe, 1968; Jaffe *et al.*, 1973). Detailed studies on bean α-amylase inhibitors began in 1975 (Marshall and Lauda, 1975) and the activity in this area has intensified in recent years. In addition, α-amylase inhibitors have also been reported in acorns (Stankovic and Markovic, 1960–1961, 1962), maize (Blanco-Labra and Iturbe-Chiñas, 1981), barley (Munday *et al.*, 1984), sorghum (Kutty and Pattabiraman, 1986a,b), millet (ragi) (Shivaraj and Pattabiraman, 1980), rye (Granum, 1978; Manjunath *et al.*, 1983) and bajra (Chandrasekher and Pattabiraman, 1985), mangoes (Mattoo and Modi, 1970), peanuts (Irshad and Sharma, 1981) and taro roots (Narayana Rao *et al.*, 1967, 1970).

In 1972, α-amylase inhibitors were reported in microorganisms (Frommer *et al.*, 1972). More than 100 publications are now available in this area alone. Some of the genes of α-amylase inhibitors have been cloned into microorganisms (Koller *et al.*, 1984; Saito *et al.*, 1986; Koller and Riess, 1987) and molecular structures and amino-acid sequences are known. Some attention has focused on the mechanism of recognition, binding and inhibition of α-amylases. There are only two reports of α-amylase inhibitors in animal tissues (Fossum and Whitaker, 1974; Zvyagintseva *et al.*, 1982a,b), excluding antibodies against α-amylases.

How widespread α-amylase inhibitors are in plants, animals and micro-organisms is unknown. Perhaps they will be found to be as ubiquitous as the trypsin-type protease inhibitors. Several double-headed protease/α-amylase inhibitors are known (Manjunath *et al.*, 1983; Mundy *et al.*, 1984; Svendsen *et al.*, 1986; Tashiro and Maki, 1986) and will be described in this chapter.

Investigations of naturally occurring α-amylase inhibitors are important because of their potential effect on human nutrition (Savaiano *et al.*, 1977; Bo-Linn *et al.*, 1982; Carlson *et al.*, 1983; Granum *et al.*, 1983; Lajolo *et al.*, 1984; Liener *et al.*, 1984; Layer *et al*, 1985), control of insects and plant regulation (Richardson, 1977; Yetter *et al.*, 1979; Powers and Culbertson, 1982; Huesing *et al.*, 1991), treatment of diabetes and hyperglycemia (Puls and Keup, 1973), analytical uses (Harmoinen *et al.*, 1986; Courtois and Franckson, 1985; Buono-core *et al.*, 1984) and as model systems for protein–protein interactions.

The three types of naturally occurring α-amylase inhibitors, their structures and their interactions with α-amylases will be described in this chapter. The reader is referred also to several other reviews on these α-amylase inhibitors (Whitaker and Feeney, 1973; Whitaker, 1981, 1983, 1989; Richardson, 1981; Garcia-Olmedo *et al.*, 1987).

5.2 Types of α-amylase inhibitors

The α-amylase inhibitors can be divided into different types on the basis of several criteria. Since this chapter is devoted to naturally occurring α-amylase inhibitors they shall be divided into types based on their chemical nature, namely (i) the microbial N-containing carbohydrates; (ii) the microbial polypeptides; and (iii) the higher plant proteins.

5.2.1 N-containing carbohydrates

Since α-amylases recognize carbohydrates such as starches and glycogens as substrates, it is not surprising that derived carbohydrates are inhibitors. Maltose is a poor competitive inhibitor, with a K_m near 0.05 M for porcine pancreatic α-amylase. Larger fragments formed from polysaccharide hydrolysis compete with starch and glycogen in binding to the enzyme, thereby functioning as competitive inhibitors and substrates simultaneously, since only one compound can be bound at the active site of the enzyme at a given time. The basic N-containing carbohydrates, produced by various *Streptomyces* species, are particularly effective inhibitors. There are three types of these inhibitors: the oligostatins, the amylostatins and the trestatins (Figure 5.1). All have in common a pseudodisaccharide unit, oligobioamine or dehydro-oligobioamine unit, a variable number of α-D-glucose units linked α-1,4 and (in one type) α-1,1. The oligobioamine unit is composed of a saturated or unsaturated cyclitol unit, with the same configuration (orientation) of side-chain constituents as found in an α-D-glucose unit, covalently attached to the amino group of 4,6-dideoxy-4-amino-D-glucopyranose. The three oligostatins differ one from the other in the number of α-1,4 glucopyranose units attached to one or both ends of the saturated oligo-bioamine unit (see Figure 5.1). The amylostatins contain a dehydro-oligobioamine unit but differ in the number of α-1,4 glucopyranose units on either end of the dehydro-oligobioamine unit. The trestatins contain one to three dehydro-oligobioamine units and terminate with one residue of α-D-glucopyranose linked 1,1- to the preceding glucose unit. Therefore, unlike the oligostatins and the amylostatins, the trestatins are nonreducing carbohydrates.

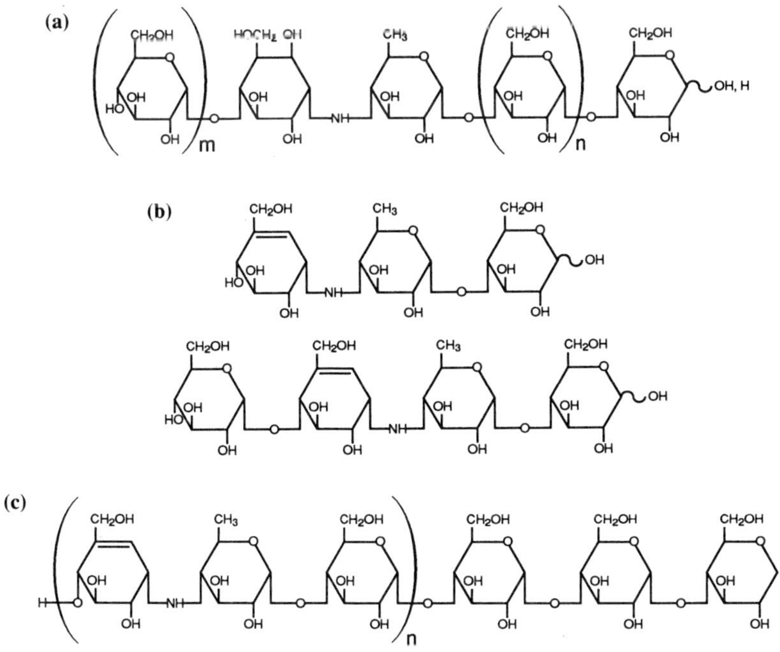

Figure 5.1 Structures of the three types of N-containing carbohydrate α-amylase inhibitors: (a) oligo-statins, C : m = 0, n = 2; D : m = 0, n = 3; E : m = 1, n = 3 (b) amylostatins and (c) trestatins, A : n = 2; B : n = 1; C : n = 3.

The N-containing carbohydrates inhibit several glucosidases, including α-amylase. Owing to the importance of these compounds in modulating effects of diabetes and hyperglycemia, substantial research has been carried out on the effect of their structure on inhibition of glycosidases. The oligostatins inhibit α-amylases and have antibiotic activity against Gram-negative bacteria (Omoto et al., 1981). The amylostatins inhibit endo- and exo-type α-amylases as well as invertase (Fukuhara et al., 1982). The endo-type α-amylases are more strongly inhibited by GXG, GXGG or GXGGG (where X = dehydro-oligobioamine; G = glucose) and the exo-type α-amylases and invertase are more strongly inhibited by XG and XGG (Schmidt et al., 1977). XGG (designated BAYg 5421) is the strongest inhibitor of invertase, while GGXGG and GGGXGG are the strongest inhibitors of animal and human α-amylases (Schmidt et al., 1977).

5.2.2 Microbial-derived polypeptide α-amylase inhibitors

Tajiri et al. (1983) reported 14 kinds of amylase inhibitors from cultures of *Streptomyces* species no. 80. Some were N-containing carbohydrates, others were polypeptides. Polypeptides of 3936–8500 Daltons with human α-amylase inhibitory properties are produced by several species of *Streptomyces* (Table 5.1). They form 1:1 complexes with human α-amylase in most cases but two molecules of Paim bind to one α-amylase molecule (Arai et al., 1985a). They are composed of a single polypeptide chain and do not contain any carbohydrate residues.

There is considerable sequence homology among the five polypeptide inhibitors (Figure 5.2). Of particular importance are the sequence –Trp–Arg–Tyr– at positions 18–20 and residues 59–70 in binding to α-amylase (Arai et al., 1985a,b; Hirayama et al., 1987). The tertiary structure of Hoe-467A, determined by X-ray crystallography, shows the –Trp–Arg–Tyr– sequence as a particularly prominent projection on the surface of the molecule and that residues 59–70 are

Table 5.1 Some chemical and physical properties of microbial peptide α-amylase inhibitors[a]

Name	Source	Molecular weight (Daltons)	Subunits No.	Carbohydrate content	Stoichiometry (amylase:inhibitor)
Paim[b] I	*Streptomyces corchorushii*	7420(73)[c]	1	0	1:2
	Streptomyces sp. No. 291–10	~6000			
Z-2685	*S. parvullus* FH-1641	8129(76)[c]	1	0	
AI-3688	*S. aureofaciens*	3936	1	0	
Haim II[d]	*S. griseosporus*	~8500(77)[c]	1	0	1:1
Hoe-467A	*S. tendae* 4158 (Tendamistat)	7958(74)[c]	1	0	1:1
AI-B	*S. viridosporus* No. 297-A2	8000			
X-2	*S. fradiae*	6500		0	

[a]Adapted from Whitaker (1989).
[b]Paim, pig pancreatic α-amylase inhibitor from microbe.
[c]Number of amino acid residues per mol.
[d]Haim I has 75 amino acids; Haim, human pancreatic α-amylase inhibitor of microbial origin.

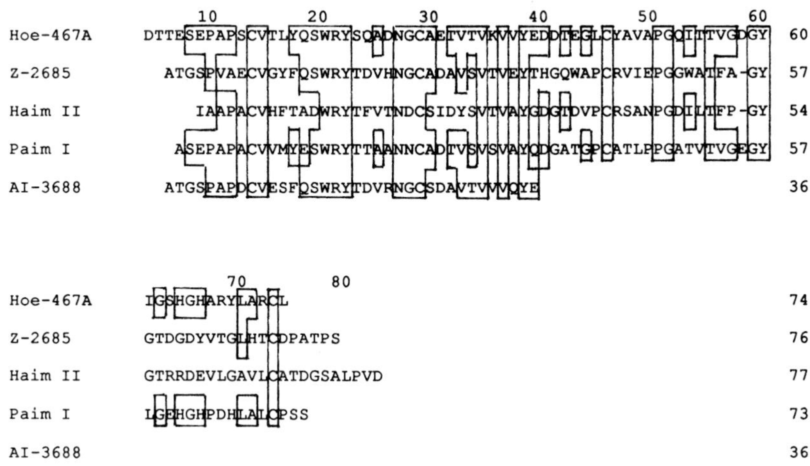

Figure 5.2 Comparison of the amino acid sequences of the *Streptomyces* polypeptide α-amylase inhibitors Hoe-467A, Z-2685, Haim II, Paim I (Hirayama *et al.*, 1987) and AI-3688 (Vertesy and Tripier, 1985). The boxed segments indicate strict homology. A dash – in the amino-acid sequence is used for aligning,

aligned in proximity to the –Trp–Arg–Tyr– sequence. The four 1/2-Cys residues at positions 11, 27, 45 and 73 are located at identical sequences in the inhibitors as is Val at positions 12, 31, 33, and 35, Asn at position 25, Tyr at positions 37 and 60, Gly at positions 51 and 59, and Thr at position 55.

The microbial-derived polypeptide α-amylase inhibitors do not appear to have been tested with plant, insect and most animal α-amylases.

5.2.3 Plant-derived protein α-amylase inhibitors

The α-amylase inhibitors have been demonstrated in several higher plants as indicated in the Introduction (see also Table 5.2). Wheat, barley, millet and common beans (*Phaseolus vulgarus*) contain several different inhibitors, some of which are isoinhibitors and others which have quite different properties, including inhibitory specificity (Table 5.3).

5.2.3.1 Inhibitors from wheat. Extensive research has been performed on the α-amylase inhibitors of wheat. There are three classes, based on molecular weight differences and three types based on electrophoretic mobility. The type 0.28 variants are monomeric, with molecular weights near 12 000 Daltons. They are more active against insect α-amylases than against human salivary or pancreatic α-amylases. The type 0.19 variant inhibitors have molecular weights around 24 000 Daltons and can be dissociated into two identical subunits (Maeda *et al.*, 1985) of about 12 000 Daltons. They are more active against insect α-amylases. The type 0.28 inhibitors (≃12 000 Daltons molecular weight) form 2:1

Table 5.2 Chemical and physical properties of plant α-amylase inhibitors

Source	Molecular weight (Daltons)	Subunits (No.)	Carbo-hydrate (%)	pI	Stoichiometry (amylase: inhibitor)
Wheat					
1 Endogenous α-amylase inhibitor	19 641(180)[b]	1			
2. WAI-65	24 000	2			
3. 0.19[c]-inhibitor	26 500(248)[b]	2		7.1	1:1
4. 0.28[c]-inhibitor	12 000(13 400)[d]			6.2	1:2
5. 0.53[c]-inhibitor	26 390	2[j]			
6. Amylase/subtilisin inhibitor[e]	20 500			7.2	
7. Tetrameric inhibitor	63 000(48 000)[f]				
Barley					
1. Amylase inhibitor	20 000			7.3	
2. Amylase/subtilisin inhibitor	19 685(181)[b]	1	0		
3. Tetrameric inhibitor	~ 60 000[g]				
Millet					
1. Foxtail	19 000			10	
2. Proso	14 000				
3. α-amylase inhibitor 1–2	9 333(95)[b]	1			
4. Amylase/trypsin inhibitor	13 300(122)[b]	1		>10	
5. Amylase/trypsin inhibitor	12 000				
6. *Echinocloa fruneutacea* grain (millet)	15 000				
Sorghum	21 000	2			
Corn	29 600	1	0		
Peanut	25 000		0		1:2
Bean					
Red kidney	49 000	4	8.6;13.0		1:1
Black kidney bean I	49 000 (55 000)[h]	3	Yes	4.93	1:1
Black kidney bean II	47 000 (60 000)		Yes	4.86	1:1
White kidney bean	38 000[h]	2	9–10		1:1
Pinto	45 000		14	4.68	
Taro root[i]	11 900	1	0		

[a]Adapted from Whitaker (1989).
[b]Number of amino acids per mol.
[c]So named based on their electrophoretic motility relative to bromophenol blue.
[d]Kashlan and Richardson, 1981.
[e]May be same as the wheat endogenous α-amylase inhibitor (see 1 above).
[f]63 000 Daltons (O'Connor & McGeeney, 1981a); 48 000 Daltons (Buonocore *et al.*, 1985).
[g]Sanchez-Monge *et al.*, 1987.
[h]Unpublished data.
[i]Seltzer and Strumeyer, 1990.
[j]Identical.

inhibitor:enzyme (I:E) complexes with α-amylases, while the type 0.19 inhibitors (≃ 24 000 Daltons) form 1:1 complexes with α-amylases. Both types contain one residue of carbohydrate per subunit; it has been suggested but not proven that the carbohydrate residue may be important in the inhibition of α-amylase (Silano *et*

Table 5.3 Specificity of various α-amylase inhibitors

Source	Designation (if any)	α-Amylases				Gluco-amylase	Other gluco-hydrolases
		Animal	Insects	Plant	Microbial		
Wheat	1. Bifunctional α-amylase/ subtilisin inhibitor	−[b]		+	−	−	−
	2. α-Amylase inhibitors	+	+	−	+[b,c]		
Barley	1. Bifunctional α-amylase subtilisin inhibitor			+			
	2. CMa		+				
Corn		−	+	+[d]	+[e]		
Sorghum		+[f]			−		
Rye		+		−	+[g]		
Millet	1. Bifunctional α-amylase/+ trypsin inhibitor						
	2. Amylase inhibitor	+[h]		−			
	3. *Echinocloa fruneutacea* grain	+[i]		+[i]			
Bajra		+					
Beans	1. Red kidney bean	+	+	−	−		
Peanuts		+		−	−		
Streptomyces							
	1. Haim	+[j]					
	2. Paim	+[k]					
	3. S-AI[l]	+				+	−
	4. S-GI[l]	+[m]			−[m]		
	5. BAYe4609 and BAYg5241	+				+	
Streptomyces calidus	NRRL 8141	+					+
Streptomyces nigrifacien	NTU-3314	+				+	

[a]Adapted from Whitaker (1989).

[b]−, no inhibition found; +, inhibition found; blank, data unknown.

[c]On a *Bacillus subtilis* α-amylase of 93 000 Daltons.

[d]Inhibitory to corn α-amylase only; not wheat, rye, barley, or triticale.

[e]Inhibitory to *B. subtilis* α-amylase, not *Aspergillus oryzae* α-amylase.

[f]Strong inhibition of human salivary and pancreatic α-amylases, bovine pancreatic α-amylase and porcine pancreatic α-amylase (less anionic form); no inhibition of more anionic porcine pancreatic α-amylase.

[g]Thermanyl (NOVO bacterial α-amylase).

[h]100 times more effective on human pancreatic α-amylase than human salivary α-amylase; no inhibition of porcine pancreatic α-amylase inhibitor.

[i]27:6:1:1 ratio for inhibition of human pancreatic, *Aspergillus oryzae*, human salivary and porcine pancreatic α-amylases.

[j]Strongly inhibits α-amylases from vertebrates; weakly inhibits α-amylases from invertebrates.

[k]Does not inhibit human and rabbit salivary and pancreatic α-amylases.

[l]From *Streptomyces diastaticum amylostaticus* and *S. lavendulae*, respectively.

[m]Inhibits animal invertases, not microbial invertases.

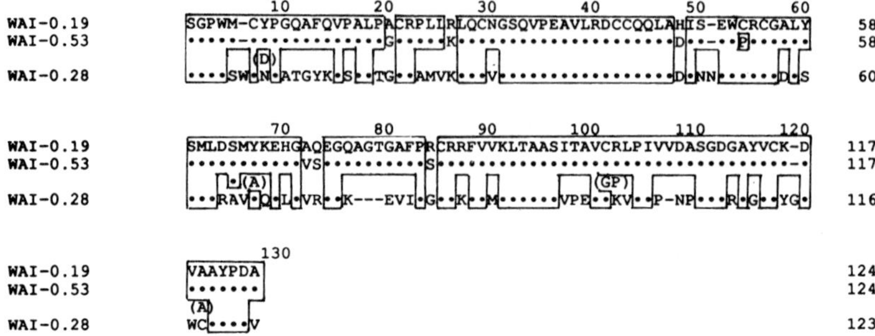

Figure 5.3 Amino-acid sequence homology among the wheat α-amylase inhibitors (WAI) 0.19, 0.53, and 0.28 (Maeda *et al.*, 1985). The boxed segments, as well as •, indicate strict homology; – used for alignment. WAI-0.28 has several isoinhibitors, as shown by changed amino-acid residues above the main sequence.

al., 1977; Silano and Zahnley, 1978). The type 0.53 variant is very similar in amino-acid sequence to type 0.19 (Figure 5.3).

The properties of the tetrameric wheat α-amylase inhibitor are not well determined. An inhibitor of 63 000 Daltons was isolated by O'Connor and McGeeney (1981a); it could be dissociated into subunits of 14 000 and 15 000 Daltons, with no carbohydrate residues. Buonocore *et al.* (1985) reported a molecular weight of 48 000 Daltons by equilibrium sedimentation and obtained four electrophoretically different bands, all of which were inhibitory. The tetrameric enzyme inhibited both insect and human α-amylases but not the enzymes from bacteria and fungi (O'Connor *et al.*, 1981b). More recent data indicate the tetrameric inhibitor is composed of proteins CM2, CM3 and CM16 from tetraploid wheat and CM2, CM3, CM16 and CM1 and CM17 from hexaploid wheat (Sanchez-Monge *et al.*, 1986). These proteins have similar properties to those of the monomeric and dimeric inhibitors, leading DePonte *et al.* (1976) to suggest that all three classes of inhibitors arose from a common ancestral protein of 12 000 Daltons (Figure 5.4). The major amino-acid sequence homology among inhibitor types 0.19, 0.28 and 0.53 is shown in Figure 5.3 and Table 5.4. There is appreciable amino-acid sequence homology among wheat α-amylase inhibitor type 0.28, ragi α-amylase/trypsin inhibitor and barley trypsin inhibitor (Figure 5.5; Table 5.4).

A double-headed α-amylase/subtilisin inhibitor has been described also in wheat (Table 5.3).

5.2.3.2 Inhibitors from barley. At least three proteins with α-amylase inhibitory activity have been isolated from barley: an α-amylase inhibitor of 20 000 Daltons (Table 5.3; Weselake *et al.*, 1983); an α-amylase/subtilisin inhibitor of 19 685 Daltons (Table 5.3; Svendsen *et al.*, 1986) and an α-amylase/protease inhibitor of approximately 13 000 Daltons (Mundy and Rogers, 1986). The barley α-amylase/ subtilisin inhibitor (BASI) has substantial amino-acid sequence homology with the Kunitz soybean trypsin inhibitor (SBTI) and the Winged bean trypsin inhibitor

Table 5.4 Primary sequence homology (%) among several cereal α-amylase inhibitors[a]

	BASI[b]	*SBTI*		*RATI*	*BTI*
SBTI	40		*BTI*	66	
WBTI	43	56	*WAI*	39	27
	RAI	*BAPI*		*WAI–0.19*	*WAI–0.53*
BAPI	62		*WAI–0.53*	96	
MATI	35	30	*WAI–0.28*	62	52

[a]Identical plus conserved homology.
[b]The acronyms used are: BAPI, barley α-amylase/protease inhibitor; BASI, barley α-amylase/subtilisin inhibitor; BTI, barley trypsin inhibitor; MATI, maize α-amylase/trypsin inhibitor; RAI, ragi α-amylase inhibitor; RATI, ragi α-amylase/trypsin inhibitor; SBTI, Kunitz soybean trypsin inhibitor; WAI, wheat α-amylase inhibitor;

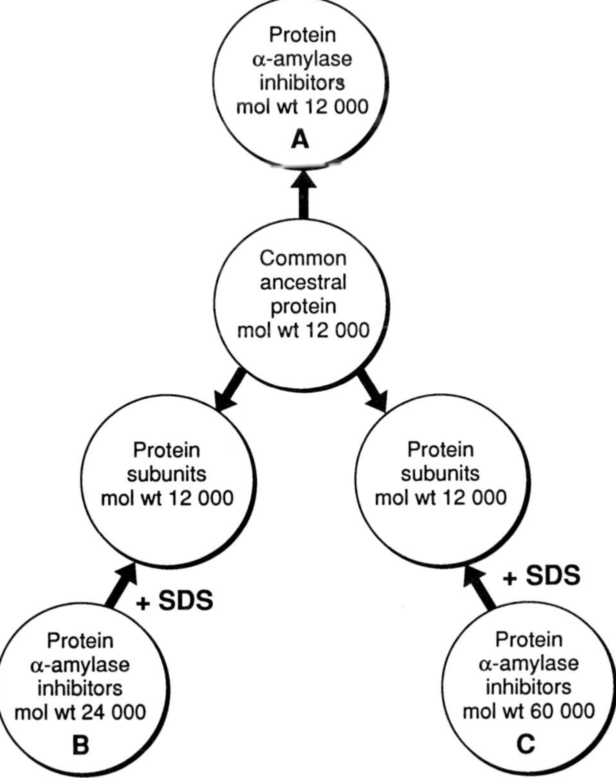

Figure 5.4 A possible pathway for evolution of the wheat monomeric (≃ 12 000 Daltons), dimeric (≃ 24 000 Daltons) and tetrameric (≃ 60 000 Daltons) α-amylase inhibitors from a common ancestral gene (from DePonte *et al.*, 1976 and Vittozzi and Silano, 1976).

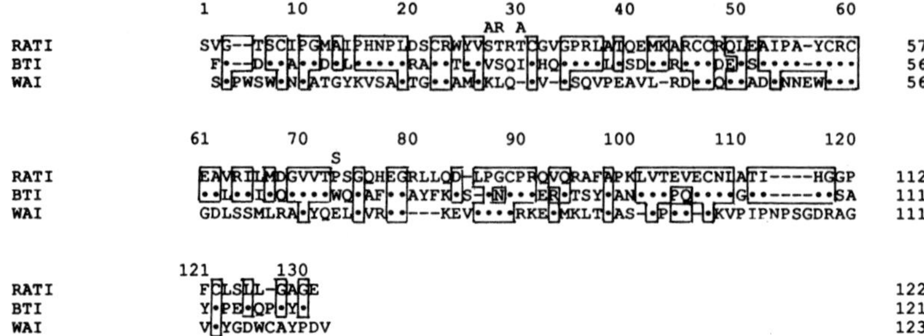

Figure 5.5 Amino-acid sequence homology among ragi α-amylase/trypsin inhibitor (RATI), barley trypsin inhibitor (BTI) and wheat α-amylase inhibitor (WAI 0.28) (from Odani *et al.*, 1983). See legend of Figure 5.3 for additional explanation.

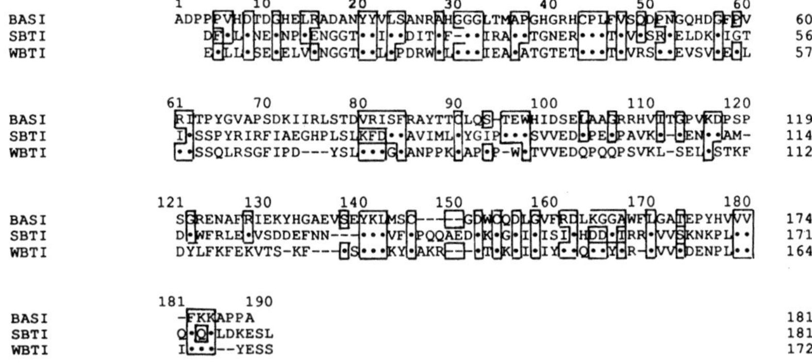

Figure 5.6 Amino-acid sequence homology among barley α-amylase/subtilisin inhibitor (BASI), Kunitz soybean trypsin inhibitor (SBTI) and winged bean trypsin inhibitor (WBTI) (from Svendsen *et al.*, 1986). See legend of Figure 5.3 for additional explanation.

(WBTI) (Figure 5.6) while the barley α-amylase/protease inhibitor (BAPI) has little homology with BASI but considerable amino-acid sequence homology with ragi α-amylase inhibitor (RAI) (Campos and Richardson, 1984) and maize α-amylase/trypsin inhibitor (MATI) (Figure 5.7).

There are three classes of barley α-amylase inhibitors, based on molecular weight; the monomeric, dimeric and tetrameric molecules are very similar to those of wheat. The tetrameric inhibitor is composed of barley proteins CMa, CMb and CMd (Sanchez-Monge *et al.*, 1986). Protein CMa is the only one active as an α-amylase inhibitor alone (Barber *et al.*, 1986a,b; Sanchez-Monge *et al.*, 1986) and is the monomeric inhibitor. Binary mixtures of any two of proteins CMa, CMb and CMd are active as α-amylase inhibitors; however, a mixture of the three proteins gives the highest specific activity (Sanchez-Monge *et al.*,

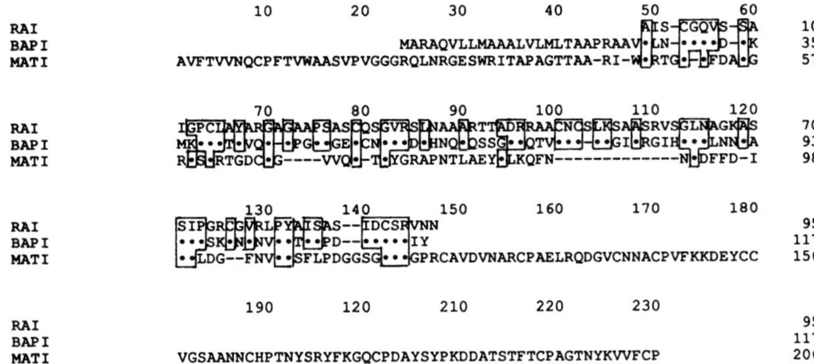

Figure 5.7 Amino-acid sequence homology among ragi α-amylase inhibitor (RAI), barley α-amylase/protease inhibitor (BAPI) and maize α-amylase/trypsin inhibitor (MATI) (from Richardson *et al.*, 1987). See legend of Figure 5.3 for additional explanation.

1986). The tetrameric inhibitor is active against *Tenebrio molitor* α-amylase but not against human salivary α-amylase. Similar subunitted proteins are found in a tetrameric inhibitor of wild barley (Sanchez-Monge *et al.*, 1987).

In summary, it appears that the barley α-amylase inhibitors evolved from a common ancestral gene coding for a protein of about 12 000 Daltons, similar to the wheat α-amylase inhibitors (see Figure 5.4).

5.2.3.3 Inhibitors from millet (ragi). Millet contains several α-amylase inhibitors, of which at least two are double-headed α-amylase/trypsin inhibitors (Table 5.2; Shivaraj and Pattabiraman, 1980). Unlike the other α-amylase inhibitors listed in Table 5.2, at least two of the millet α-amylase inhibitors have pI values near 10. The ragi α-amylase/trypsin inhibitor (RATI) (Campos and Richardson, 1983) has substantial amino-acid sequence homology with a barley trypsin inhibitor (BTI) and a wheat α-amylase inhibitor (WAI 0.28) (Figure 5.5). There is some sequence homology between RAI, BAPI and MATI (Figure 5.7), although the latter two proteins are larger than RAI.

The data in Table 5.2 indicate there may be two classes of α-amylase inhibitors based on molecular weight, those with molecular weight near 12 000 Daltons and those with a molecular weight of 19 000 Daltons.

5.2.3.4 Inhibitors from maize. Two double-headed α-amylase/trypsin inhibitors of molecular weight near 29 600 (Blanco-Labra and Iturbe-Chiñas, 1981) and 22 077 Daltons (Richardson *et al.*, 1987) have been isolated from maize (Table 5.2). The 29 600 Dalton protein is somewhat unique in that it inhibits maize and bacterial and insect α-amylases but not human α-amylases. The amino-acid sequence of the 22 077 Dalton protein has been determined (Richardson *et al.*,

1987; Figure 5.7); it has considerable amino-acid sequence homology with BAPI and RAI, although the maize inhibitor is much larger than the other two proteins.

5.2.3.5 Other inhibitors from cereals. Another model for some α-amylase inhibitors appears to be heterodimeric proteins in which the two subunits are covalently linked by disulfide bonds. This type of inhibitor was reported in rye (Granum, 1978), pearl millet (Chandrasekher and Pattabiraman, 1985), *Echinocloa frunentacea* (Kutty and Pattabiraman, 1985) and sorghum (Kutty and Pattabiraman, 1986a,b).

As indicated by the data in Table 5.5, the cereal α-amylase inhibitors all contain appreciable amounts of 1/2-Cys residues, ranging from 2.21% for BASI to 8.94% for WAI 0.28. As shown in Figures 5.3 and 5.5–5.7, the positions of the 1/2-Cys residues are strictly conserved among the α-amylase inhibitors within a group and between homologous α-amylase inhibitors and protease inhibitors, as well as in double-headed α-amylase/protease inhibitors.

Essentially nothing is known about the mechanism of inhibition of α-amylase by the cereal α-amylase inhibitors. It has been postulated that the single carbo-hydrate residue per subunit of inhibitor in some α-amylase inhibitors may be involved in inhibition (Silano *et al.*, 1977; Silano and Zahnley, 1978); however, not all cereal α-amylase inhibitors contain carbohydrate.

5.2.3.6 Common bean (Phaseolus vulgaris) *α-amylase inhibitors.* In 79 of 95 varieties of legumes tested α-amylase inhibitory activity was reported (Jaffe *et al.*, 1973). Inhibitors have been purified from black, white and red kidney beans

Table 5.5   1/2-Cys content of α-amylase and protease inhibitors from cereals[a]

Inhibitor	1/2-Cys (mol %)
Barley α-amylase/subtilisin inhibitor (BASI)	2.21
Kunitz soy bean trypsin inhibitor (SBTI)	2.21
Winged-bean trypsin inhibitor (WBTI)	2.33
Ragi α-amylase inhibitor (RAI)	7.37
Barley α-amylase/protease inhibitor (BAPI)	6.84
Maize α-amylase/trypsin inhibitor (MATI)	7.77
Ragi α-amylase/trypsin inhibitor (RATI)	8.20
Barley trypsin inhibitor (BTI)	8.26
Wheat α-amylase inhibitor (WAI–0.28)	8.94
Wheat α-amylase inhibitors	
(WAI–0.19)	8.06
(WAI–0.53)	7.26
(WAI–0.28)	8.94

[a]From amino-acid sequences shown in Figures 5.3 and 5.5–5.7.

and pinto beans (Marshall and Lauda, 1975; Powers and Whitaker, 1977a,b; Frels and Rupnow, 1984; Wilcox and Whitaker, 1984b; Lajolo and Finardi Filho, 1985; Kotaru *et al.*, 1987; Ho and Whitaker, 1993a,b; X. Yin and J.R. Whitaker, 1994, unpublished data).

5.2.3.7 Molecular weights. By gel filtration chromatography, the molecular weight of the inhibitors ranges from 40 000 Daltons for the white kidney bean inhibitor (Ho and Whitaker, 1993a,b) to 60 000 Daltons for inhibitor I-1 from black beans (X. Yin and J.R. Whitaker, 1994, unpublished data) (Table 5.6). By polyacrylamide gel electrophoresis (PAGE) (7%, 8%, 9% and 10% gels), the molecular weights ranged from 38 000 Daltons for the white kidney bean inhibitor (Ho and Whitaker, 1993a,b) to 60 000 Daltons for inhibitor I-1 from black beans (Yin and Whitaker, 1994, unpublished data). Although the two methods gave similar results, the estimated molecular weights are too high owing to the substantial carbohydrate content of all the inhibitors. These estimates may be as much as 8% to 10% too high.

5.2.3.8 Isoelectric points. The inhibitor proteins are all acidic, with pI values ranging from 4.35 for black bean var. Rico 23 (Lajolo and Finardi Filho, 1985) to 5.0 for white kidney bean No. 858A (Ho and Whitaker, 1993a,b) (Table 5.6). The cDNA for the inhibitor from *P. vulgaris* cv. Greensleeves has been sequenced (Moreno and Chrispeels, 1989). The data show 29 Asp and 16 Glu residues and 5 His, 13 Arg and 18 Lys (extrapolated to 45 000 Dalton protein). An approximate calculation gives a pI of 4.6 (assuming an average pK_a of 4.0 for the carboxyl groups).

5.2.3.9 Carbohydrate content. The carbohydrate content of several of the inhibitors (Table 5.6) ranges from 7.5% for black bean I-1 (Frels and Rupnow, 1984) to 14.5% for black bean var. Rico 23 (Lajolo and Finardi Filho, 1985). Two different cultivars of red kidney beans had 8.6% (Powers and Whitaker, 1977a) and 13.0% (Wilcox and Whitaker, 1984b) carbohydrate. Isoinhibitors from the same black bean contained from 7.5–9.0% carbohydrate (Frels and Rupnow, 1984).

A 6 M urea-PAGE of the black bean inhibitors I-1 and I-2 (X. Yin and J.R. Whitaker, 1994, unpublished data), black bean var. Rico 23 (F. Finardi Filho and F. Lajolo, 1994, unpublished data) and white kidney bean No. 858A inhibitor (Ho and Whitaker, 1993a,b) are shown in Figure 5.8. The four inhibitors are quite similar. At the top are two major sets of bands, some smaller bands and then a series of closely spaced bands at the bottom. These closely spaced bands only separate in urea-containing gels. When the inhibitor is deglycosylated, a single protein band is obtained from these closely spaced bands. The authors believe the multiple small bands are due to different extents of glycosylation, while the larger bands are due to polypeptide subunits.

Red kidney bean inhibitor, with 13.0% carbohydrate, contained 25 mannose, 2 xylose, 1 fucose and 17 N-acetylglucosamine residues/mol (Wilcox and Whitaker, 1984b). All subunits of the inhibitor contained at least one glyco group. The cDNA-determined sequence of the Greensleeves bean indicated a minimum of five glycosylation sites in the approximately 25 000 Dalton unit sequence, proposed as the principal subunit of the inhibitor (Moreno and Chrispeels, 1989).

5.2.3.10 1/2-Cystine and methionine contents. The α-amylase inhibitors have few 1/2-Cys residues (Tables 5.6 and 5.7). The black bean var. Rico 23 has one cysteine residue as determined by amino-acid analysis (Lajolo and Finardi Filho, 1985) and by titration (Whitaker *et al.*, 1988) with Ellman's reagent (5,5'-dithiobis-(2-nitrobenzoic acid); Ellman, 1958). The residue is buried and only available to Ellman's reagent after denaturation (heat plus 1% SDS; Whitaker *et al.*, 1988). Frels and Rupnow (1984) reported two and one 1/2-Cys residues in black bean inhibitors I-1 and I-2, respectively. Powers and Whitaker (1977a) reported four 1/2-Cys residues in red kidney bean inhibitor. They are sulfhydryl groups, not disulfides. The cDNA-determined sequence of Greensleeves beans does not include cysteine/cystine (Moreno and Chrispeels, 1989). No 1/2-Cys residues were found in black bean isoinhibitors I-1 and I-2 (Yin and Whitaker, 1994, unpublished data), white kidney bean inhibitor (Ho and Whitaker, 1993a,b) and pinto bean (Kotaru *et al.*, 1987).

The methionine content ranged from no residues for the inhibitor from black bean I-2 (Yin and Whitaker, 1994, unpublished data) to six residues for black bean I-2 (Frels and Rupnow, 1984) (Table 5.6). Greensleeves beans contain one Met residue, based on the cDNA sequence (Moreno and Chrispeels, 1989).

5.2.3.11 N-terminal amino acids. The N-terminal amino acid(s) have been determined for four inhibitors. Black bean var. Rico 23 inhibitor contained Ala, Glu and Thr as N-terminal residues as determined by the dansyl chloride method (Lajolo and Finardi Filho, 1985). More recently, using the DABITC method (Chang, 1983), Ala and Ser have been identified as the N-terminal amino acids for black bean var. Rico 23 inhibitor (F. Finardi Filho and F. Lajolo, 1994, unpublished data) and black bean inhibitor I-1 (X. Yin and J.R. Whitaker, 1994, unpublished data). A white kidney bean inhibitor 858A gave only Ala as the N-terminal residue (Ho and Whitaker, 1993a,b), consistent with other evidence that it is composed of two identical subunits. The cDNA-determined sequence for the inhibitor from Greensleeves beans indicated that Ala and Ser were N-terminal, although the possibility for 2 Ala residues exists (Moreno and Chrispeels, 1989).

5.2.3.12 Subunit composition. There is no doubt that all the α-amylase inhibitors studied so far contain two or more subunits, based on comparison of PAGE and SDS-PAGE results. Powers and Whitaker (1977a) first found that SDS-PAGE gave at least three subunits of 15 000–17 000, 12 000–15 000 and

Table 5.6 Some chemical and physical characteristics of bean α-amylase inhibitors

Parameter	Red kidney bean[a]	Black beans					White bean		Pinto bean[g]
		Rico 23[b]	I-1[c]	I-2[c]	I-1[d]	I-2[d]	e	f	
MW (kD)	49[h]	56[h]; 53[h] 49[i]	60[h] 60[i]	49[h] 55[i]	49[h]	47[h]	40[h]; 42[h] 38[i]	49[h]	45
pI	—	4.35	4.75	4.65	4.93	4.86	5.0	—	4.68
Carbohydrate (%)[j]	8.6[a]; 13.0[k]	14.5; 8.4[l]	9.9	10.2	7.5	9.0	11.0	9–10	14.0
1/2 Cys (residues/MW)	4	1	0	0	2	1	0	0	0
N-terminal AA	—	Ala, Glu, Thr[m]; Ala, Ser[n]		Ala, Ser[n]	Ala, Ser[n]	—	—	Ala[n]	—
Subunits (No.)	3–4	3	3–4	3–4	2	3	2	—	—
Subunit MW (kD)[o]	15–17 12–15 11–12	15–17.5	13–17.5 23[p]	13–17.5	15.5–17 32–34	15–17	18–20[q]	—	—

[a]Powers and Whitaker (1977a).
[b]Lajolo and Finardi Filho (1985).
[c]Yin, X. and Whitaker, J.R. (1994 unpublished data).
[d]Frels and Rupnow (1984).
[e]Ho and Whitaker (1993a,b); white kidney bean No. 858A.
[f]Marshall and Lauda (1975).
[g]Kotaru et al. (1987).
[h]By gel filtration.
[i]Polyacrylamide gel electrophoresis (PAGE).
[j]Phenol-sulfuric acid method (Dubois et al., 1956).
[k]Wilcox and Whitaker (1984b).
[l]From composition after hydrolysis.
[m]Dansyl chloride method (Gray, 1972).
[n]DABITC method (Chang, 1983).
[o]SDS-PAGE.
[p]By SDS-PAGE when β-mercaptoethanol is not included.
[q]Identical subunits.

Table 5.7   1/2-Cys content of α-amylase inhibitors from legumes and tubers[a]

Inhibitor	1/2-Cys (mol %)
Legumes	
Black Bean FI-2 (Rico 23)	0.26
Greensleeves	0.00
Black Bean YI-1	0.00
Black Bean YI-2	0.00
White Kidney Bean	0.00
Tuber	
Esculentamin	0.00

[a]From amino-acid sequence data of proteins in Figures 5.12 and 5.13.

11 000–12 000 Daltons for a red kidney bean α-amylase inhibitor. Three protein bands are clearly indicated, with the middle band staining about twice more intensely than the two outside bands. Similar results have been found for other inhibitors. Black bean var. Rico 23 is thought to have three subunits of 15 000–17 000 Daltons, in part based on finding Ala, Glu and Thr as N-terminal residues (Lajolo and Finardi Filho, 1985) and the presence of three electrophoretically distinct bands. Black bean inhibitors I-1 and I-2 appear to be composed of three or four subunits of 13 000–17 000 Daltons, although a subunit of 23 000 Daltons is found when β-mercaptoethanol is omitted from SDS-PAGE (Yin and Whitaker, 1994, unpublished data). Inhibitors I-1 and I-2 of another black bean variety appear to have two and three subunits, respectively (Frels and Rupnow, 1984). Inhibitor I-1 gave two subunits of 15 500–17 000 and 32 000–34 000 Daltons, respectively, by SDS-PAGE. Inhibitor I-2 gave three subunits of 15 000–17 000 Daltons. White kidney bean 858A has two subunits of 18 000–20 000 Daltons that are probably identical (Ho and Whitaker, 1993a,b). Difficulties in separating and determining the subunits are as follows: (i) they are very hydrophobic; (ii) they are not very soluble when denatured; and (iii) minor bands of varying molecular weight are always present (possibly protease-produced fragments), no matter how homogeneous the inhibitor appears.

Moreno and Chrispeels (1989) proposed that the Greensleeves bean inhibitor consists of two subunits of approximately 8700 and 16 500 Daltons, based on the cDNA-determined sequence (Figure 5.9) and its similarity to a lectin-like protein (Hoffman *et al.*, 1982). The proposed molecular weight of approximately 27 700 Daltons (25 200 Daltons for protein plus 2500 Daltons carbohydrate arbitrarily taken as 10%) is smaller than any of the inhibitors found and may not be the molecular weight of the mature inhibitor. Ho and Whitaker (1993a,b) have recently isolated an inhibitor from a white bean with a molecular weight of 20 000–24 000 Daltons composed of 9000 and 15 000 Dalton subunits. The subunits reported in other inhibitors do not quite fit this scheme (Table 5.6). The

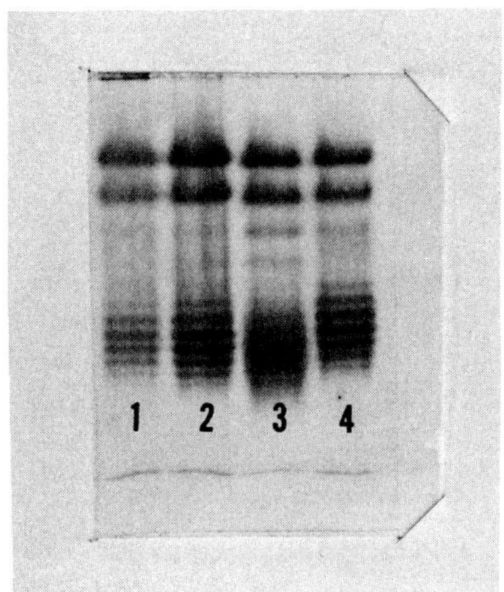

Figure 5.8 PAGE electrophoresis of four common bean α-amylase inhibitors. The inhibitors were treated with 6 M urea at pH 8.9 and then ≃20 μg inhibitor was loaded on the gel prepared in the same solvent. The proteins were fixed and stained with Coomassie brilliant blue R-250 dye. 1, black bean inhibitor I-1 (from Yin and Whitaker, 1994, unpublished data); 2, black bean inhibitor I 2 (from Yin and Whitaker, 1994, unpublished data); 3, black bean var. Rico 23 inhibitor (from Finardi Filho and Lajolo, 1994, unpublished data); white kidney bean 858A inhibitor (from Ho and Whitaker, 1993a,b).

23 000 Dalton subunit from black bean inhibitor I-1 (Yin and Whitaker, 1994, unpublished data; Table 5.6) could possibly be related to the 27 700 Dalton unit proposed by Moreno and Chrispeels (1989).

Based on our present level of understanding of the biosynthesis and maturation of the bean α-amylase inhibitors and their subunit diversity, we propose the present model as a working hypothesis (Figure 5.10). The preproinhibitor (precursor) is similar to the lectin-like protein reported by Hoffman *et al.* (1982) of 27 800 Dalton (Figures 5.9 and 5.10). The precursor is composed of a signal peptide unit of 2600 Daltons and a proinhibitor protein of 25 200 Daltons. The signal peptide is removed by a specific protease that cleaves the peptide bond between Ser_{23} and Ala_{24} (Figure 5.9). Another specific protease may cleave between Asn_{100} and Ser_{101} to give 8700 and 16 500 Dalton subunits (Figure 5.9). These two subunits then assemble, primarily by hydrophobic bonding, to yield the five different inhibitors proposed in Figure 5.10. For reasons not yet known, all possible combinations are not found in a single bean. The authors have found two isoinhibitors in a black bean variety that can now be explained by this model (BB-YI-1 of approximately 55 000 Daltons and BB-YI-2 of approximately 60 000 Daltons; Yin and Whitaker, 1994, unpublished data). The complexity of the system is further compounded by specific differences among the proposed subunit molecular weights from different varieties of beans (Figure 5.10). Further support for this model is provided by amino-acid composition and amino-acid sequence data for these inhibitors.

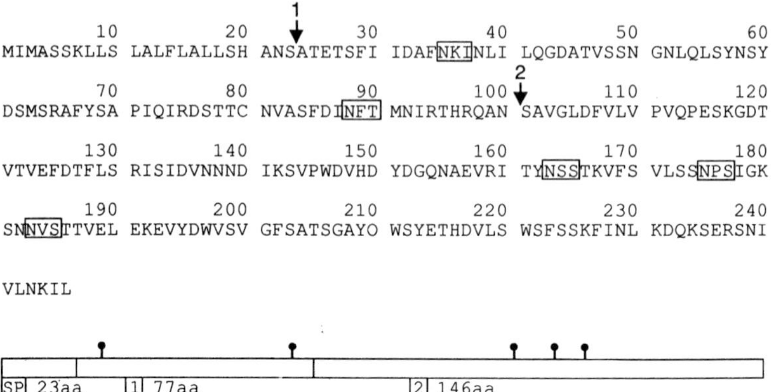

Figure 5.9 Comparison of the amino-acid sequence of lectin-like protein (LLP) deduced from the nucleotide sequence of the cloned gene (from Hoffman *et al.*, 1982) with the partial sequence of Greensleeves bean α-amylase inhibitor (underlined segmênts of 11 and 15 N-terminal amino-acid sequences obtained for the α-amylase inhibitor subunits). The vertical arrow indicates the signal peptide cleavage site deduced by homology with lectins. The bottom diagram indicates the proposed processing of LLP involving signal peptide removal, glycosylation and cleavage of the mature polypeptide into two fragments. The boxed tripeptide amino acids regions indicate five potential sites for glycosylation (from Moreno and Chrispeels, 1989).

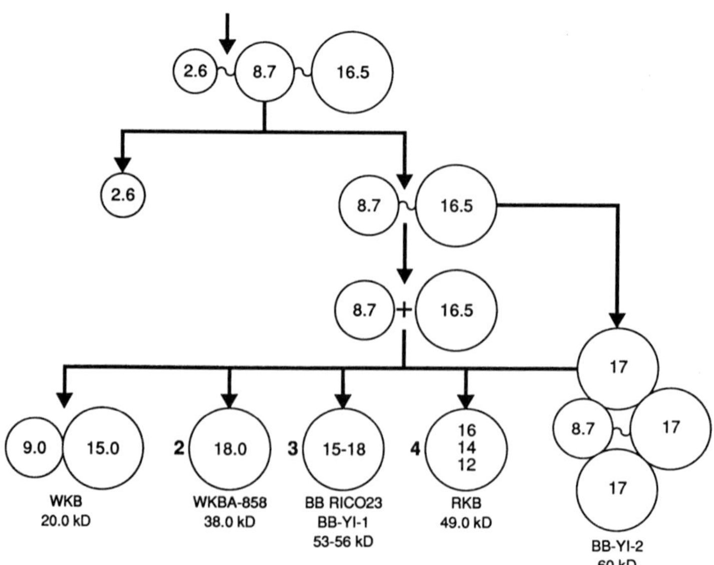

Figure 5.10 Hypothetical model showing the possible derivation of six different α-amylase inhibitors from a preprotein similar to LLP (see Figure 5.9).

5.2.3.13 Amino-acid composition. The amino-acid composition of nine inhibitors from seven different varieties/cultivars of beans are given in Table 5.8. The results come from six different laboratories and by two different techniques (hydrolysis/amino-acid analyser and by cDNA sequence mapping). As can be seen in Table 5.8, the amino-acid compositions are remarkably similar, with a few notable exceptions, when all the molecular weights are normalized to 50 000 Daltons. The exceptions include 1/2-Cys values, which range from zero in five cases to one in the case of black bean var. Rico 23 (Lajolo and Finardi Filho, 1985), two and one in black bean I-1 and I-2, respectively (Frels and Rupnow, 1984), and four in red kidney bean (Powers and Whitaker, 1977a) inhibitors. Valine was markedly higher in the white kidney bean inhibitor (Ho and Whitaker, 1993a,b) than in the other inhibitors. Methionine was not detected in the black bean inhibitor I-2 (Yin and Whitaker, 1994, unpublished data), while it ranged from three to six residues per 50 000 Daltons in the other inhibitors. Tyrosine content of black bean inhibitor I-2 (Yin and Whitaker, 1994, unpublished data) and of white kidney bean inhibitor (Ho and Whitaker, 1993a,b) were about 60% that of the other inhibitors. The His content of black bean var. Rico 23 (Lajolo and Finardi Filho, 1985) was two, while there were 10 and 14 residues in white kidney bean inhibitor (Ho and Whitaker, 1993a,b) and black bean inhibitor I-1 (Frels and Rupnow, 1984), respectively, compared with an average of six His residues in the other inhibitors.

The last column of Table 5.8 shows the amino acid composition for lectin-like protein (LLP) based on the sequence of the cDNA that encodes the protein. The only statistically significant difference between the average ±2 SD of eight inhibitors and the cDNA-determined amino-acid composition for LLP is for Ile. The Ile composition for LLP was 25 versus 17 ± 1 for the other eight inhibitors. There is no doubt that Moreno and Chrispeels (1989) have located the gene for the α-amylase inhibitor in the Greensleeves bean variety and have cloned and expressed it in *Escherichia coli.*

Of special interest is the average 57%:43% nonpolar:polar ratio of amino acids of the inhibitors. This explains why the subunits associate to form the mature inhibitors and why they are so difficult to dissociate. Identifying the gene and cloning it into *E. coli* will undoubtedly accelerate work on the structures of the inhibitors.

5.2.3.14 Tryptic peptide maps. Tryptic peptide maps for black bean inhibitor var. Rico 23 (Finardi Filho and Lajolo, 1994, unpublished data), black bean inhibitors I-1 and I-2 (Yin and Whitaker, 1994, unpublished data) and white kidney bean 858A inhibitor (Ho and Whitaker, 1993a,b) are shown in Figure 5.11. There is a remarkable similarity among the peptides, perhaps even more than one would expect since several of the peptides are glycosylated. The number of peptides obtained are less than half that expected from the combined Lys and Arg contents of the compounds. The expected results based on amino-acid data (Table 5.8; with maximum number observed) are: black bean I-1 inhibitor, 33(9); black

Table 5.8 Amino acid composition of several bean α-amylase inhibitors[a]

Amino acid	Red kidney bean[b]	Black bean					White kidney bean No. A858[f]	Pinto bean[g]	Average ±SD	LLP[h]
		Rico 23[c]	I-1[d]	I-2[d]	I-1[e]	I-2[e]				
Asx (Asp+Asn)	62	73	60	69	56	64	57	67	64 ± 6	65[j]
Thr	33	30	35	37	30	31	28	34	32 ± 3	32
Ser	53	49	51	55	46	49	58	54	52 ± 4	59
Gix (Glu+Gln)	35	34	33	33	34	36	32	29	33 ± 2	30[j]
Pro	11	12	14	9	14	12	11	9	12 ± 2	9
Gly	21	19	25	20	29	25	16	18	22 ± 4	16
Ala	25	22	27	23	27	26	13[i]	22	23 ± 5 (25 ± 2)[i]	20
1/2-Cys	4	1	0	0	2	1	0	0	1 ± 1	0
Val	35	34	34	35	34	34	41[i]	38	36 ± 3 (35 ± 1)[i]	38
Met	4	3	3	0[i]	5	6	3	3	3 ± 2 (4 ± 1)[i]	4
Ile	17	17	18	19	17	19	14	17	17 ± 2	25
Leu	17	19	25	24	25	25	20	19	22 ± 3	23
Tyr	12	15	15	8[i]	13	12	8[i]	14	12 ± 3 (14 ± 1)[i]	14
Phe	19	22	24	25	22	21	18	21	22 ± 2	21
His	6	2[i]	7	6	14[i]	7	9	5	7 ± 3 (7 ± 1)[i]	5
Lys	20	14	19	18	23	19	13[i]	18	18 ± 3	18
Arg	13	12	13	15	18	15	8	11	13 ± 3	13
Trp	—	2	—	—	—	—	10	—		7

[a]Normalized to 50 000 Daltons.
[b]Wilcox and Whitaker (1984b); 45 000 Daltons; 13.0% carbohydrate.
[c]Lajolo and Finardi Filho (1985); 53 000 Daltons; 14.5% carbohydrate.
[d]Yin and Whitaker (1994 unpublished data); I-1, 60 000 Daltons; 9/9% carbohydrate. I-2, 55 000 Daltons; 10.2% carbohydrate
[e]Frels and Rupnow (1984); I-1, 49 000 Daltons; 7.5% carbohydrate. I-2, 47 000 Daltons; 9.0% carbohydrate.
[f]Ho and Whitaker (1993a,b) 38 000 Daltons; 11% carbohydrate.
[g]Kotaru et al. (1987); 45 000 Daltons; 14.0% carbohydrate.
[h]Moreno and Chrispeels (1989); assumed 10% carbohydrate in calculations; LLP, lectin-like protein.
[i]Data not included in average (>2 SD).
[j]29 Asp and 36 Asn; 16 Glu and 14 Gln.

bean I-2 inhibitor, 34(11), black bean var. Rico 23 inhibitor, 27(10); white kidney bean inhibitor, 18(8). While substantially lower observed values may represent peptide bonds not hydrolyzed, it seems more reasonable to suggest that they provide additional evidence for the presence of two to four subunits of similar amino-acid sequences.

5.2.3.15 Stability of inhibitors to subunit dissociation. The stabilities of the black bean var. Rico 23 inhibitor, the black bean inhibitors I-1 and I-2 and white

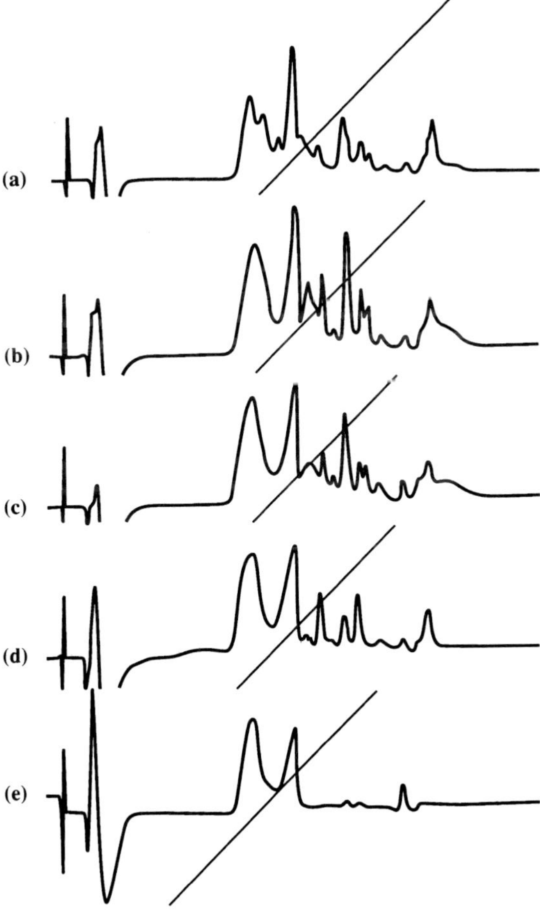

Figure 5.11 Tryptic peptide maps for four common bean α-amylase inhibitors. The peptides were separated by FPLC chromatography on a Mono Q anion-exchange column. Digestion conditions: trypsin:inhibitor ratio 1:200, 5 h at pH 8.0 and 25°C. (a) black bean inhibitor I-1 (from Yin and Whitaker, 1994, unpublished data); (b) black bean inhibitor I-2 (from Yin and Whitaker, 1994, unpublished data); (c) black bean var. Rico 23 inhibitor (from Finardi Filho and Lajolo, 1994, unpublished data); (d) white kidney bean 858A inhibitor (from Ho and Whitaker, 1993a,b); (e) buffer control.

kidney bean 858A inhibitor to complete dissociation in 2% SDS plus 10% β-mercaptoethanol at pH 8.8 were different (Finardi Filho and Whitaker, 1994, unpublished data). White kidney bean inhibitor dissociated within 3 min at 60°C (it was approximately 60% dissociated at 0°C); black bean var. Rico 23 dissociated within 3 min at 80°C (it was approximately 40% dissociated at 0°C); black bean inhibitor I-1 required <3 min at 40°C (it was >90% dissociated at 0°C) and black bean inhibitor I-2 required about 3 min at 60°C (it was approximately 25% dissociated at 0°C). At 100°C all the inhibitors were completely dissociated within 30 s. The extent of dissociation was determined using 12.5% PAGE.

5.2.3.16 Amino-acid sequences of the bean α-amylase inhibitors.   While none of the bean α-amylase inhibitors has been completely sequenced, there are sufficient data available to develop a working model and to draw some conclusions. The available data are shown in Figure 5.12. The starting point for this alignment of the α-amylase inhibitor sequences is based on the results of Moreno and Chrispeels (1989) showing that the partial sequence for the Greensleeves α-amylase inhibitor has 100% homology with the lectin-like protein (LLP) of Hoffman *et al.* (1982). Osborn *et al.* (1988) showed that there is substantial amino-acid sequence homology among bean proteins LLP, arcelin-1 (from a wild bean), lectin E and lectin L. Figure 5.12 shows the partial amino-acid sequence of a white kidney bean inhibitor (Ho and Whitaker, 1993a,b), the two black bean isoinhibitors BB-YI-1 and BB-YI-2 (Yin and Whitaker, 1994, unpublished data) and one of the two black bean Rico 23 isoinhibitors BB-FI-2 (Finardi Filho and Lajolo, 1994, unpublished data). Since mature α-amylase inhibitors were used, no signal peptide sequence is shown. However, from these data, it is clear that all these proteins have evolved from a common ancestral gene. The data in Table 5.9 demonstrate the identical and conserved amino-acid sequence homology among the nine proteins. Of particular note, all the proteins are post-translationally processed (hydrolyzed) between Ser_{23} and Ala_{24} (or Ser_{24} in two cases) and seven of the proteins are known to be further processed at Ser_{107}.

While the amino-acid sequence data are incomplete for the α-amylase inhibitors and some sequences cannot be aligned, the evidence is overwhelming that the four groups of proteins have indeed evolved from a common ancestral gene. To date, researchers have been working independently, but to gain a better understanding of these proteins they must cooperate and maximize their future efforts. To begin with, there is a need to determine whether any of the other three proteins inhibit α-amylases, which proteins are toxic to insects, and to obtain more complete amino-acid sequence data for these inhibitors. In the longer term, the biosynthesis, maturation and tertiary structures of these proteins need to be understood in order to understand their differences in mechanism of action. Obviously, many more questions than answers are presently available.

5.2.3.17 Esculentamin.   Taro root contains the α-amylase inhibitor, esculentamin, of 11 900 Daltons (Seltzer and Strumeyer, 1990). A partial amino-acid

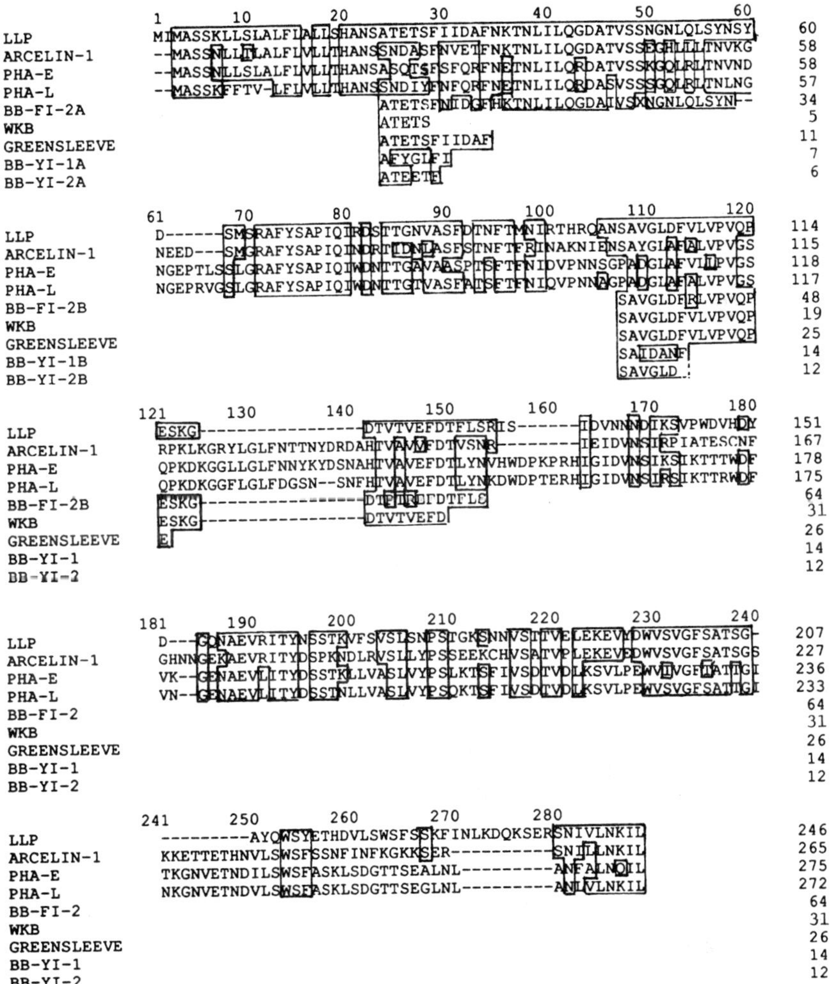

Figure 5.12 Amino-acid sequence homology among several common bean (*Phaseolus vulgaris*) proteins: lectin-like protein (LLP) (from Hoffman *et al.*, 1982); arcelin–1 (Osborn *et al.*, 1988); (PHA-E and PHA-L) phytohemagglutinins-E and -L (from Hoffman and Donaldson, 1985); BB-FI-2A and 2B, black bean var. Rico 23 α-amylase isoinhibitor 2, subunits A and B (from Finardi Filho and Lajolo, 1994, unpublished data), white kidney bean (WKB) (from Ho and Whitaker, 1993a,b, unpublished data); white common bean cv. Greensleeves (from Moreno and Chrispeels, 1989); BB-YI-1A and 1B, black bean α-amylase isoinhibitor 1, subunits A and B (from Yin and Whitaker, 1994, unpublished data); BB-YI-2A and 2B, black bean α-amylase isoinhibitor 2, subunits A and B (from Yin and Whitaker, 1994, unpublished data). The boxed segments show strictly conserved homology among all proteins. The – is used in alignment.

Table 5.9 Amino-acid sequence homology (%) among several bean-binding proteins[a]

	LLP[b]	Arcelin-1	PHA-E	PHA-L	BB-FI-2	Greensleeves	BB-YI-1	BB-YI-2
Arcelin-1	63							
PHA-E	61	59						
PHA-L	61	63	88					
BB-FI-2[c]	88	53	58	55				
Greensleeves[c]	100	54	62	46	92			
BB-YI-1[c]	57	43	36	36	50	57		
BB-YI-2[c]	92	58	58	50	92	50	33	
WKB[c]	100	42	50	35	88	100	33	82

[a]Identical plus conserved homology.
[b]Acronyms used are: LLP, lectin-like protein; PHA, phytohemagglutinin; BB-FI, black bean var. Rico 23-Finardi inhibitor; BB-YI, black bean-Yin inhibitor; WKB, white kidney bean.
[c]Incomplete sequence.

sequence has been determined, as shown in Figure 5.13. There is no sequence homology between esculentamin and any of the sequenced α-amylase inhibitors described in this review. Like the legume α-amylase inhibitors, it does not contain 1/2-Cys residues (Table 5.7).

5.2.3.18 Mechanistic aspects of the action of bean α-amylase inhibitors. The bean (*Phaseolus vulgaris*) α-amylase inhibitors investigated so far form 1:1 stoichiometric complexes with porcine pancreatic and human salivary α-amylases. Pick and Wöber (1978) reported that white bean (unspecified variety) inhibitor formed a 1:1 complex with salivary α-amylase but the gel filtration chromatography results were consistent with a 2:1 (amylase/inhibitor) complex. Microbial and higher plant α-amylases are not inhibited by the *Phaseolus vulgaris* inhibitors.

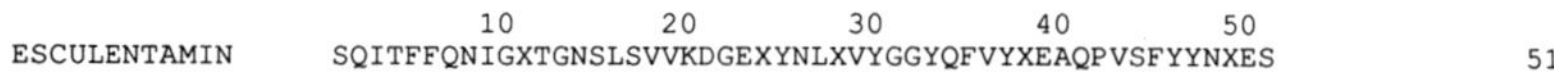

Figure 5.13 Partial amino-acid sequence of α-amylase inhibitor esculentamin from taro root (from Seltzer and Strumeyer, 1990).

Little information is available on how the α-amylase and α-amylase inhibitors recognize each other and how the enzyme becomes inhibited in the process. Inactivation of the enzyme is very slow, requiring 30–120 min to reach equilibrium depending on the experimental conditions (Powers and Whitaker, 1977b). The rate of complex formation is at least 20 times more rapid at pH 5.4 than at pH 6.9 (Powers and Whitaker, 1977b; Tanizaki and Lajolo, 1985; Whitaker *et al.*, 1988). The reaction is second order at pH 6.9 and first order at pH 5.4 (Tanizaki and Lajolo, 1985; Whitaker *et al.*, 1988). The complex formed between black bean var. Rico 23 inhibitor and porcine pancreatic α-amylase had a dissociation constant (K_d) of 1.7×10^{-10} M at pH 5.5 and 4.4×10^{-9} M at pH

6.9 (Tanizaki and Lajolo, 1985). The K_d for red kidney bean inhibitor and porcine pancreatic α-amylase at pH 6.9 was 3.5×10^{-11} M (Powers and Whitaker, 1977b) while that for the white kidney bean 858A inhibitor was 1.0×10^{-11} M (Ho and Whitaker, 1993a,b). Calcium-free porcine pancreatic α-amylase binds to black bean var. Rico 23 inhibitor at the same rate at pH 5.4 and 6.9 as with the Ca^{2+}-containing enzyme (Whitaker et al., 1988). Chloride ions are required for binding at pH 6.9 but not at pH 5.4 (Whitaker et al., 1988).

Wilcox and Whitaker (1984a) proposed that red kidney bean inhibitor and bovine pancreatic α-amylase rapidly (almost diffusion controlled) form an initial complex enzyme inhibitor (EI) that is fully active; with time EI changes to a complex, EI*, with approximately 5% of the original activity on p-nitrophenyl α-maltoside but with no activity on starch (equation 5.1)

$$E \; + \; I \; \underset{k_{-1}}{\overset{k_1}{\rightleftharpoons}} \; EI \; \underset{k_{-2}}{\overset{k_2}{\rightleftharpoons}} \; EI^* \tag{5.1}$$

The equilibrium constant ($K_{eq} = k_{-1}/k_1$) for the first step was 3.5×10^{-5} M, while the equilibrium constant ($K_{eq} = k_{-2}/k_2$) for the second step was 3.5×10^{-11} M. Additional data and kinetic modelling led to the reactions shown in Figure 5.14. Figure 5.14 is fully consistent with all experimental data. The rate constant, k_2, was determined to be 3.05 min^{-1}, while $k_{-2} \simeq 0$. Assuming a value of about 1×10^6 M for k_1, k_{-1} would be about 31 s^{-1}. The data indicate that E, EI and EI* all bind with substrate. With a small substrate, such as p-nitrophenyl α-maltoside, ES and EIS form products at the same rate, while EI*S hydrolyzes p-nitrophenyl α-maltoside at about 5% this rate. Starch is not hydrolyzed but still binds to the complex, as does maltose (Powers and Whitaker, 1977b).

The effect on complex formation of modifying groups on either the α-amylase inhibitor or the α-amylase have been investigated. Wilcox and Whitaker (1984b) reported that enzymatic removal of at least 70% of the glycosyl groups from the red kidney bean inhibitor did not cause activity loss. The removed glycosyl groups alone at 3.3×10^5 M excess relative to the α-amylase concentration did not cause activity loss. Oxidation of one Trp residue with N-bromosuccinimide caused 50% loss of activity. Modification of three of the five His residues with diethylpyrocarbonate caused about 50% loss in activity. Treatment with periodate caused loss of two Tyr and one Met residues per mol of inhibitor; there was no direct correlation between oxidation of these two residues and activity loss. Similar results were observed for the white kidney bean 858A inhibitor (Ho and Whitaker, 1993a,b).

Several modifications of porcine pancreatic α-amylase and their effect on binding of the modified enzyme to black kidney bean var. Rico 23 inhibitor have been

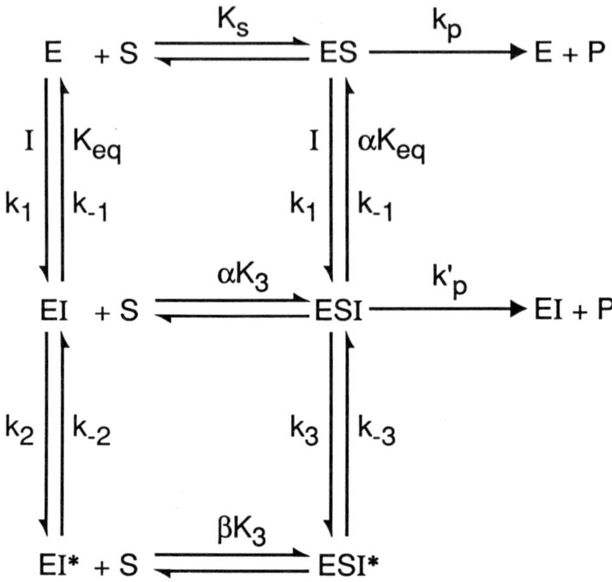

Figure 5.14 Proposed pathway for complex formation of red kidney bean α-amylase inhibitor with porcine pancreatic α-amylase in presence of substrate, *p*-nitrophenyl maltoside (from Wilcox and Whitaker, 1984a).

reported (Tanizaki *et al.*, 1985; Whitaker *et al.*, 1988). Modification of His residues of α-amylase with rose bengal or derivatization with diethylpyrocarbonate, Trp residues with N-bromosuccinimide and Tyr residues with N-acetylimidazole, all reduced α-amylase activity; however, none of the modifications affected complex formation with the inhibitor. Whitaker *et al.* (1988) showed that modification of the single –SH group of cysteine in black kidney bean var. Rico 23 inhibitor or the –SH groups of the two cysteine residues of porcine pancreatic α-amylase had no effect on complex formation. Calcium-free α-amylase bound inhibitor at both pH 5.4 and 6.9 as well as the Ca^{2+}-containing enzyme. Chloride ions, essential for porcine pancreatic α-amylase activity, were required for binding the inhibitor at pH 6.9, but not at pH 5.4.

Much work is needed to determine the mechanisms of recognition and binding between the α-amylase inhibitors and α-amylases.

References

Arai, M., Oouchi, N. and Murao, S. (1985a) Inhibitory properties of an α-amylase inhibitor, Paim, from *Streptomyces corchorushii. Agric. Biol. Chem.* **49**:987–91.

Arai, M., Oouchi, N., Goto, A., Ogura, S. and Murao, A. (1985b) Chemical modification of Haim, a proteinaceous α-amylase inhibitor. *Agric. Biol. Chem.* **49**:1523–24.

Barber, D., Sanchez-Monge, R., Garcia-Olmedo, F., Salcedo, G. and Mendez, E. (1986a) Evolutionary implications of sequential homologies among members of the trypsin/α-amylase inhibitor family (CM-proteins) in wheat and barley. *Biochim. Biophys. Acta* 873:147–51.

Barber, D., Sanchez-Monge, R., Mendez, E., Lazaro, A., Garcia-Olmedo, F. and Salcedo, G. (1986b) New α-amylase and trypsin inhibitors among the CM-proteins of barley (*Hordeum vulgare*). *Biochim. Biophys. Acta* **869**:115–18.

Blanco-Labra, A. and Iturbe-Chiñas, F.A. (1981) Purification and characterization of an α-amylase inhibitor from maize (*Zea maize*) *J. Food Biochem.* **5**:1–17.

Bo-Linn, G.W., Santa Ana, C.A., Morawski, S.G. and Fordtran, J.S. (1982) Starch blockers—their effect on calorie absorption from a high starch meal. *New Engl. J. Med.* **307**:1413–16.

Bowman, D.E. (1945) Amylase inhibitor of navy beans. *Science* **102**:358–59.

Buonocore, V., Giardina, P., Parlamenti, R., Poerio, E. and Silano, V. (1984) Characterization of chicken pancreas α-amylase isozymes and interaction with protein inhibitors from wheat kernel. *J. Sci. Food Agric.* **35**:225–32.

Buonocore, V., de Biasi, M.-G., Giardina, P., Poerio, E. and Silano, V. (1985) Purification and properties of an α-amylase tetrameric inhibitor from wheat kernel. *Biochim. Biophys. Acta* **831**:40–48.

Campos, F.A.P. and Richardson, M. (1983) The complete amino-acid sequence of the bifunctional α-amylase/trypsin inhibitor from seeds of ragi (Indian finger millet, *Elevsine coracana Gaertn.*) *FEBS Lett.* **152**:300–4.

Campos, F.A.P. and Richardson, M. (1984) The complete amino-acid sequence of the α-amylase inhibitor I-2 from seeds of ragi (Indian finger millet, *Elevsine coracana Gaertn.*) *FEBS Lett.* **167**:221–25.

Carlson, G.I., Li, B.U.K., Bass, P. and Olsen, W.A. (1983) A bean α-amylase inhibitor formulation (starch blocker) is ineffective in man. *Science* **219**:393–95.

Chandrasekher, G. and Pattabiraman, T.N. (1985) Natural plant enzyme inhibitors. XIX. Purification and properties of an α-amylase inhibitor from bajra (*Peunisetum typhoideum*). *Indian J. Biochem. Biophys.* **20**:241–45.

Chang, J.-Y. (1983) Amino-terminal analysis with dimethylaminoazobenzene isothiocyanate (DABITC). *Methods Enzymol.* **91**:79–84.

Chrzaszcz, T. and Janicki, J. (1933) 'Sisto-amylase', ein naturlicher paralysator der amylase. *Biochem. Z.* **260**:354–68.

Chrzaszcz, T. and Janicki, J. (1934) XII. The inactivation of animal amylase by plant paralysers and the presence of inactivating substances in solutions of animal amylases. *Biochem. J.* **28**:296–304.

Courtois, P. and Franckson, J.R.M. (1985) Evaluation of a new procedure for isoamylase measurement by selective inhibition. *J. Clin. Chem. Clin. Biochem.* **23**:733–37.

DePonte, R., Parlamenti, R., Petrucci, T., Silano, V. and Tomasi, M. (1976) Albumin α-amylase inhibitors from wheat flour. *Cereal Chem.* **53**:805–20.

Dubois, M., Gilles, K.A., Hamilton, J.K., Rebers, P.A. and Smith, F. (1956) Colorimetric method for determination of sugars and related substrates. *Anal. Chem.* **28**:350–56.

Ellman, G.L. (1958) A colorimetric method for determining low levels of mercaptans. *Arch. Biochem. Biophys.* **74**:443–50.

Fossum, K. and Whitaker, J.R. (1974) Simple method for detecting amylase inhibitors in biological materials. *J. Nutr.* **104**:930–36.

Frels, J.M. and Rupnow, J.H. (1984) Purification and partial characterization of two α-amylase inhibitors from black bean (*Phaseolus vulgaris*). *J. Food Biochem.* **8**:281–301.

Frommer, W., Puls, W., Schaefer, D. and Schmidt, D. (1972) Glycoside hydrolase inhibitors from actinomycetes. *Ger. Offen*, DE 2/064/092.

Fukuhara, K., Murai, H. and Murao, S. (1982) Amylostatins, other amylase inhibitors produced by *Streptomyces diastaticus* subspecies amylostaticus No. 2476. *Agric. Biol. Chem.* **46**:2021–31.

Garcia-Olmedo, F., Salcedo, G., Sanchez-Monge, R., Gomez, L., Royo, J. and Carbonero, P. (1987) Plant proteinaceous inhibitors of proteinases and α-amylases. *Oxford Surveys of Plant Mol. Cell Biol.* **4**:275–334.

Granum, P.E. (1978) Purification and characterization of an α-amylase inhibitor from rye (*Secale cereale*) flour. *J. Food Biochem.* **2**:103–20.

Granum, P.E., Holm, H., Wilcox, E. and Whitaker, J.R. (1983) Inhibitory properties of two commercially available starch blocker preparations. *Nutr. Repts. Intl.* **28**:1233–44.

Gray, W.R. (1972) End-group analysis using dansyl chloride. *Methods Enzymol.* **25**:121–38.

Harmoinen, A., Jokela, H., Koivula, T. and Poppe, W. (1986) Evaluation of a new inhibitor test for isoamylase on Hitachi 705 analyzer. *J. Clin. Chem. Clin. Biochem.* **24**:903–5.

Hernandez, A. and Jaffé, W.G. (1968) Inhibitor of pancreatic amylase from beans (*Phaseolus vulgaris*). *Acta Cient. Venez.* **19**:183–85.

Hirayama, K., Takahashi, R., Akashi, S., Fukuhara, K., Oouchi, N., Murai, A., Arai, M., Murao, S., Tanaka, K. and Nojima, I. (1987) Primary structure of Paim I, an α-amylase inhibitor from *Streptomyces corchorushii*, determined by the combination of Edman degradation and fast atom bombardment and mass spectrometry. *Biochem.* **26**:6483–88.

Ho, M.F. and Whitaker, J.R. (1993a) Purification and partial characterization of white kidney beans (*Phaseolus vulgaris*) α–amylase inhibitors from two experimental cultivars. *J. Food Biochem.*, **17**:15–33.

Ho, M.F. and Whitaker, J.R. (1993b) Subunit structures and essential amino acid residues of white kidney bean (*Phaseolus vulgaris*) α-amylase inhibitors. *J. Food Biochem.*, **17**:35–52.

Hoffman, L.M. and Donaldson, D.D. (1985) Characterization of two *Phaseolus vulgaris* phytohemagglutinin genes closely linked on the chromosome. *EMBO J.* **4**:883–89.

Hoffman, L.M., Ma, Y. and Barker, R.F. (1982) Molecular cloning of *Phaseolus vulgaris* lectin mRNA and use of cDNA as a probe to estimate lectin transcript levels in various tissues. *Nucl. Acids Res.* **10**:7819–28.

Huesing, J.E., Shade, R.E., Chrispeels, M.J. and Murdock, L.L. (1991) α-Amylase inhibitor, not phytohemagglutinin, explains resistance of common bean seeds to cowpea weevil. *Plant Physiol.* **96**:993–96.

Irshad, M. and Sharma, C.B. (1981). Purification and properties of an α-amylase inhibitor from *Arachis hypogaea seeds. Biochim. Biophys. Acta* **659**:326–33.

Jaffé, W.G., Moreno, R. and Wallis, V. (1973) Amylase inhibitor in legume seeds. *Nutr. Repts. Int.* **7**:169–74.

Kashlan, N. and Richardson, M. (1981) The complete amino-acid sequence of a major wheat protein inhibitor of α-amylase. *Phytochem.* **20**:1781–84.

Kneen, E. and Sandstedt, R.M. (1943) An amylase inhibitor from certain cereals. *J. Amer. Chem. Soc.* **65**:1247–52.

Kneen, E. and Sandstedt, R.M. (1946) Distribution and general properties of an amylase inhibitor in cereal. *Arch. Biochem. Biophys.* **9**:235–49.

Koller, K.P. and Riess, G.J. (1987) Tandem promoter for *Streptomyces* expression vectors. *Ger. Offen.* DE 3/536/182.

Koller, K.P., Engels, J. and Uhlmann, E. (1984) Gene amplification and over-production of the α-amylase inhibitor (Hoe 467, Tendamistat) in *Streptomyces tendae. Eur. Congr. Biotechnol.* **3**:273–78.

Kotaru, M., Saito, K., Yoshikawa, H. Ikeuchi, T. and Ibuki, F. (1987) Purification and some properties of an α-amylase inhibitor from cranberry bean (*Phaseolus vulgaris*). *Agric. Biol. Chem.* **51**:577–78.

Kutty, A.V.M. and Pattabiraman, T.N. (1985) Isolation and characterization of an α-amylase inhibitor from *Echinocloa frunentacea* grains. *Indian J. Biochem. Biophys.* **22**:155–60.

Kutty, A.V.M. and Pattabiraman, T.N. (1986a) Differential inhibition of porcine pancreatic amylase isoenzymes by the sorghum (*Sorghum bicolor*) seed amylase inhibitor. *Biochem. Arch.* **2**:203–8.

Kutty, A.V.M. and Pattabiraman, T.N. (1986b) Isolation and characterization of an amylase inhibitor from sorghum seeds, specific for human enzymes. *J. Agric. Food Chem.* **34**:552–57.

Lajolo, F.M. and Finardi Filho, F. (1985) Partial characterization of the α-amylase inhibitor of black beans (*Phaseolus vulgaris*), variety Rico 23. *J. Agric. Food Chem.* **33**:132–38.

Lajolo, F.M., Nancini-Filho, J. and Menezes, E.W. (1984) Effect of a bean (*Phaseolus vulgaris*) α-amylase inhibitor on starch utilization. *Nutr. Repts. Int.* **30**:45–54.

Layer, P., Carlson, G.L. and Dimagno, P. (1985) Partially purified white bean amylase inhibitor reduces starch digestion *in vitro* and inactivates intraduodenal amylase in humans. *Gastroenterol.* **88**:1895–1902.

Liener, I.E., Donatucci, D.A. and Tarcza, J.C. (1984) Starch blockers: a potential source of trypsin inhibitors and lectins. *Amer. J. Clin. Nutr.* **39**:196–200.

Maeda, K., Kakabayashi, S. and Matsubara, H. (1985) Complete amino-acid sequence of an α-amylase inhibitor in wheat kernel (0.19 inhibitor). *Biochim. Biophys. Acta* **828**:213–21.

Manjunath, N., Veerabhadrappa, P.S. and Virupaksha, T.K. (1983) Isolation and characterization of a trypsin inhibitor from finger millet. *Phytochem.* **22**:2349–57.

Marshall, J.J. and Lauda, C.M. (1975) Purification and properties of phaseolamin, an inhibitor of α-amylase, from the kidney bean, *Phaseolus vulgaris*. *J. Biol. Chem.* **250**:8030–37.

Mattoo, A.K. and Modi, V.V. (1970) Partial purification and properties of enzyme inhibitors from unripe mangos. *Enzymologia* **39**:237–47.

Moreno, J. and Chrispeels, M.J. (1989) A lectin gene encodes the α-amylase inhibitor of common bean. *Proc. Natl. Acad. Sci. USA* **86**:1–5.

Mundy, J., Hejgaard, J. and Svendsen, I. (1984) Characterization of a bifunctional wheat inhibitor of endogenous α-amylase and subtilisin. *FEBS Lett.* **167**:210–14.

Mundy, J. and Rogers, J.C. (1986) Selective expression of a probable α-amylase/protease inhibitor in barley aleurone cells: comparison to the barley α-amylase/subtilisin inhibitor. *Planta* **169**:51–63.

Narayana, Rao, M., Shurpalekar, K.S. and Sundaravalli, O.E. (1967) An amylase inhibitor in *Colocasia esculenta*. *Indian J. Biochem.* **4**:185.

Narayana Rao, M., Shurpalekar, K.S. and Sundaravalli, O.E. (1970) Purification and properties of an amylase inhibitor from colocasia (*Colocasia esculenta*) tubers. *Indian J. Biochem.* **7**:241–43.

O'Connor, C.M. and McGeeney, K.F. (1981a) Isolation and characterization of four inhibitors from wheat flour which display differential inhibition specificities for human salivary and human pancreatic α-amylases. *Biochim. Biophys. Acta* **658**:387–96.

O'Connor, C.M. and McGeeney, K.F. (1981b) Interaction of human α-amylases with inhibitors from wheat flour. *Biochim. Biophys. Acta* **658**:397–405.

Odani, S., Koide, T. and Ono, T. (1983) The complete amino-acid sequence of barley trypsin inhibitors. *J. Biol. Chem.* **258**:7998–8003.

Omoto, S., Itoh, J., Ogino, H., Iwamatsu, K., Nishizawa, N. and Inouye, S. (1981) Oligostatins, new antibiotics with amylase inhibitory activity. II. Structures of oligostatins C, D and E. *J. Antibiotics* **34**:1429–33.

Osborn, T.C., Alexander, D.C., Sun, S.S.M., Cardona, C. and Bliss, F.A. (1988) Insecticidal activity and lectin homology of arcelin seed protein. *Science* **240**:207–10.

Pick, K.-H. and Wöber, G. (1978) Proteinaceous α-amylase inhibitor from beans (*Phaseolus vulgaris*). Purification and partial characterization. *Hoppe-Seyler's Z. Physiol. Chem.* **359**:1371–77.

Powers, J.R. and Culbertson, J.D. (1982) *In vitro* effect of bean amylase inhibitor on insect amylases. *J. Food Protect.* **45**:655–57.

Powers, J.R. and Whitaker, J.R. (1977a) Purification and some physical properties of red kidney bean (*Phaseolus vulgaris*) α-amylase inhibitor. *J. Food Biochem.* **1**:217–38.

Powers, J.R. and Whitaker, J.R. (1977b) Effect of several experimental parameters on combination of red kidney bean (*Phaseolus vulgaris*) α-amylase inhibitor with porcine pancreatic α-amylase. *J. Food Biochem.* **1**:239–60.

Puls, W. and Keup, U. (1973). Influence of an α-amylase inhibitor (Bay d 7791) on blood glucose, serum insulin and NEFA in starch-loading tests in rats, dogs and man. *Diabetologia* **9**:97–101.

Richardson, M. (1977) The proteinase inhibitors of plants and micro-organisms. *Phytochem.* **16**:159–69.

Richardson, M., (1981). Protein inhibitors of enzymes. *Food Chem.* **6**:235–53.

Richardson, M., Valdes-Rodriquez, S. and Blanco-Labra, A. (1987) A possible function for thaumation and a TMV-induced protein suggested by homology to a maize inhibitor. *Nature* **327**:432–434.

Saito, S., Takahasi, H., Saito, H., Arai, M and Murao, S. (1986) Molecular cloning and expression in *Streptomyces lividans* of a proteinaceous α-amylase inhibitor (Haim II) gene from *Streptomyces griseosporeus*. *Biochem. Biophys. Res. Commun.* **141**:1099–103.

Sanchez-Monge, R., Gomez, L., Garcia-Olmedo, F. and Salcedo, G. (1986) A tetrameric inhibitor of insect α-amylase from barley. *FEBS Lett.* **207**:105–9.

Sanchez-Monge, R., Fernandez, J.A. and Salcedo, G. (1987) Subunits of tetrameric α-amylase inhibitors of *Hordeum chilense* are encoded by genes located in chromosomes 4Hch and 7Hch. *Theor. Appl. Genet.* **74**:811–16.

Saunders, R..M. and Lang, J.A. (1973) α-Amylase inhibitor in *T. aestivum*: purification and physicochemical properties. *Phytochem.* **12**:1237–41.

Savaiano, D.A., Powers, J.R., Costello, M.J., Whitaker, J.R. and Clifford, A.J. (1977) The effect of an α-amylase inhibitor on the growth rate of weanling rats. *Nutr. Repts. Intl.* **15**:443–49.

Schmidt., D.D., Frommer, W., Junge, B., Mueller, L., Wingender, W., Truscheit, E. and Schaefer, D.

(1977) α-Glucosidase inhibitors. New complex oligosaccharides of microbial origin. *Naturwissenschaften* **64**:535–36.

Seltzer, R.D. and Strumeyer, D.H. (1990) Purification and characterization of esculentamin, a proteinaceous alpha-amylase inhibitor from the taro root, *Colocasia esculenta*. *J. Food Biochem.* **14**:199–217.

Shivaraj, B. and Pattabiraman, T.N. (1980) Natural plant enzyme inhibitors. VIII. Purification and properties of two α-amylase inhibitors from ragi (*Eleusine coracana*) grains. *Indian J. Biochem. Biophys.* **17**:181–85.

Silano, V., Pocchiari, F. and Kasarda, D.D. (1973) Physical characterization of α-amylase inhibitors from wheat. *Biochim. Biophys. Acta* **317**:139–48.

Silano, V., Poerio, E. and Buonocore, V. (1977) A model for the interaction of wheat monomeric and dimeric protein inhibitors with α-amylase. *Mol. Cell. Biol.* **18**:87–91.

Silano, V. and Zahnley, J.C. (1978) Association of *Tenebrio molitor* L. α-amylase with two protein inhibitors—one monomeric, one dimeric—from wheat flour. Differential scanning calorimetric comparison of heat stabilities. *Biochim. Biophys. Acta* **533**:181–85.

Stankovic, S.C. and Markovic, N.D. (1960–61). A study of amylase inhibitors in the acorn. *Glasnik Hem. Drustva, Beograd* **25–26**:519–25.

Stankovic, S.C. and Markovic, N.D. (1963) A study of amylase inhibitors in the acorn. *Chem. Abstr.* **59**:3084d.

Svendsen, I., Hejgaard, J. and Mundy, J. (1986) Complete amino-acid sequence of the α-amylase/subtilisin inhibitor from barley. *Carlsberg Res. Commun.* **51**:43–50.

Tajiri, T., Koba, Y. and Ueda, S. (1983). Amylase inhibitors produced by *Streptomyces* sp. no. 80. *Agric. Biol. Chem.* **47**:671–79.

Tanizaki, M.M. and Lajolo, F.M. (1985) Kinetics of the interaction of pancreatic α-amylase with a kidney bean (*Phaseolus vulgaris*) α-amylase inhibitor. *J. Food Biochem.* **9**:71–89.

Tanizaki, M.M., Lajolo, F.M. and Finardi Filho, F. (1985) Combination of black bean (*Phaseolus vulgaris*) amylase inhibitor with modified pancreatic α-amylase. *J. Food Biochem.* **9**:91–104.

Tashiro, M. and Maki, Z. (1986) Isolation of protein inhibitors of pepsin, trypsin, and α-amylase in the grain of foxtail millet. *Agric. Biol. Chem.* **50**:2955–57.

Vertesy, L. and Tripier, D. (1985) Isolation and structure elucidation of an α-amylase inhibitor, AI-3688, from *Streptomyces aureofaciens*. *FEBS Lett.* **185**:187–90.

Vittozzi, L. and Silano, V. (1976) The phylogenesis of protein α-amylase inhibitors from wheat seed and the speciation of polyploid wheats. *Theor. Appl. Genet.* **48**:279–84.

Weselake, R.J., MacGregor, A.W., Hill, R.D. and Duckworth, H.W. (1983) Purification and characteristics of an endogenous α-amylase inhibitor from barley kernels. *Plant Physiol.* **73**:1008–12.

Whitaker, J.R. (1981) Naturally occurring peptide and protein inhibitors of enzymes, in *Impact of Toxicology on Food Processing*, (eds. J. Ayres and J. Kirshman), Avi Publishing, Westport, Connecticut, pp. 57–104.

Whitaker, J.R. (1983). Protease and amylase inhibitors in biological materials, in *Xenobiotics in Foods and Feeds*, (eds. J. Finley and D.E. Schwass), ACS Symposium Series 234, American Chemical Society, Washington DC, pp. 15–46.

Whitaker, J.R. (1989) α-Amylase inhibitors of higher plants and microorganisms, in *Food Proteins*, (eds. J. Kinsella and W.G. Soucie). American Oil Chemical Society, Washington DC, pp. 354–380.

Whitaker, J.R. and Feeney, R.E. (1973) Enzyme inhibitors in foods, in *Toxicants Occurring Naturally in Foods*, (ed. F.M. Strong). *Natl. Acad. Sci.*, Washington DC, pp. 276–98.

Whitaker, J.R., Finardi Filho, F. and Lajolo, F.M. (1988) Parameters involved in binding of porcine pancreatic α-amylase with black bean inhibitor: role of sulfhydryl groups, chloride, calcium, solvent composition and temperature. *Biochemie* **70**:1153–61.

Wilcox, E.R. and Whitaker, J.R. (1984a) Some aspects of the mechanism of complexation of red kidney bean α-amylase inhibitor and α-amylase. *Biochem.* **23**:1783–91.

Wilcox, E.R. and Whitaker, J.R. (1984b) Structural features of red kidney bean α-amylase inhibitor important in binding with α-amylase. *J. Food Biochem.* **8**:189–213.

Yetter, M.A., Saunders, R.M. and Boles, H.P. (1979) α-Amylase inhibitors from wheat kernels as factors in resistance to postharvest insects. *Cereal Chem.* **56**:243–44.

Zvyagintseva, T.N., Sova, V.V., Pereva, I. and Elyakova, L.A. (1982a) Amylase inhibitors of Caribbean Actinia. Proteinaceous amylase inhibitors from *Stoichactis helianthus*. *Khim. Prir. Soedin* **3**:343–49.

Zvyagintseva, T.N., Sova, V.V., Pereva, I. and Elyakova, L.A. (1982b) Amylase inhibitors of Caribbean Actinia. Proteinaceous amylase inhibitors from Stoichactis helianthus. *Chem. Abstr.* 97:89155c.

6 Application of multivariate analysis in studies of food protein functions

S. NAKAI, T. AISHIMA AND R.Y. YADA

Abstract

The most commonly used multivariate analysis (MVA) techniques in food science, that is, principal component, discriminant and cluster analyses and various combinations of these are compared using nonfood related and food protein structure-function data. Although discriminant analysis (DA) is the most efficient and useful method for classifying samples into groups, there is a risk of discarding useful information, since DA uses only that portion of data variation that is useful for the classification purpose. Therefore, DA is recommended only when the groups are clearly defined *a priori*. If one wishes to group continuous data, for example native versus denatured protein without a clear distinction between groups, principal component regression is recommended. Principal component-similarity analysis (PCS) comprises a group of sequentially applied MVA techniques proposed for classification of samples when the grouping cannot be clearly defined in advance or when only a few samples are available for classification. PCS classifies samples with minimal loss of data variation.

6.1 Introduction

Today we are facing a revolution in food analysis as a result of enormous advances in instrumental analysis, especially in the areas of spectroscopy and chromatography. Analyses that were once time-consuming, slow, tedious and labour-intensive are no longer required as we move towards complete automation of analytical techniques. Owing to enormous amounts of data produced by automated instrumental analysis, efficient data-processing techniques are becoming an absolute necessity in modern food analysis. A useful solution to this increasingly critical analytical problem of data processing lies with a group of techniques called multivariate analysis (MVA). These techniques allow quantitative evaluation of the variation of a large number of random variables simultaneously.

Recently, Aishima and Nakai (1991) reviewed critically the various chemometric techniques in flavour research, as well as many methods in MVA. In addition, Jeon (1991) reviewed MVA techniques for food research and quality assurance, citing many applications for classification and differentiation of food products, product authenticity, chemical composition, quality indices and micro-biological and sensory properties. This chapter presents a comparison of MVA methods presently used in food research. A new technique is proposed, namely principal component-similarity analysis, and its use is demonstrated using non-food- and food-related data.

6.2 Multivariate analysis (MVA) techniques

The most popular and most useful MVA techniques in food research are principal component, discriminant and cluster analyses. Although often used separately, combinations of these techniques are sometimes used to allow for a more power-ful analysis of data (e.g. discriminant analysis of principal component scores, and regression analysis of principal component scores).

6.2.1 Principal component analysis

Principal component analysis (PCA) is a data transformation technique that computes linear combinations of the original variables which can then be used to summarize the data with minimal loss of information. PCA accomplishes dimension reduction by minimizing minor variations so that major variations can be summarized in terms of newly calculated variables called principal compo-nents. Information obtained can thus be depicted on only a few two-dimensional plots of principal component scores. Principal components comprise sets of orthogonal linear co-ordinates. This technique therefore avoids multicollinearity, which reduces the reliability of regression coefficients computed from multiple regression analysis (Newell and Lee, 1981).

6.2.2 Discriminant analysis

Discriminant analysis is used to classify cases or samples according to a pre-determined variable grouping (Ennis *et al.*, 1982). Stepwise linear discriminant analysis (LDA) is performed by choosing a linear combination of variables that maximizes the F-ratio of a one-way analysis of variance. In most cases, the first few significant variables are sufficient to account for almost all of the important between-group variability. To obtain a more clearly defined classification, quadratic discriminant analysis or nonparametric discriminant analysis can be used (or attempted) on the same analytical data (Girard and Nakai, 1993). When the data do not follow the multivariate normal distribution, it is recom-mended that nonparametric discriminant analysis be applied. Application of LDA to principal component scores obtained after application of PCA to analy-tical data may improve the correctness in classification. The accuracy of classi-fication is improved by increasing the number of samples in each group; however, the separability of the data is simultaneously reduced. Discriminant analysis is a typical 'supervised' method: the groups to be classified are known in advance.

6.2.3 Principal Component Regression (PCR)

PCR is the stepwise multiple regression analysis of principal component scores obtained from PCA of the original data. This method is advantageous since predictive equations for data of a continuous nature, for example spoilage and quality of food products, can be derived along with statistical parameters.

6.2.4 Principal Component Similarity Analysis (PCS)

PCS is a method whereby pattern similarity analysis of principal component scores derived from PCA of the original data is conducted. The protocol for PCS is as follows:

Step 1: Apply PCA to the original k data for samples to derive PC scores and eigenvalues E.

Step 2: Compute independent variables V as follows:

$$P_i = E_i \left/ \sum_{i=1}^{k} E_i \right. \tag{6.1}$$

$$V_i = 100 \left(1 - \sum_{i=1}^{i} P_{i-1} \right) \tag{6.2}$$

where i is the PC number.

Step 3: Compute dependent variables Y as follows:

$$Y_i = V_i + (PC_i - PC_r) \times M \tag{6.3}$$

where r is the reference and M is an arbitrary multiplier (usually $M = 10$).

Step 4: Carry out linear regression analysis for Y versus V to compute correlation coefficient (r) and slope (s).

Step 5: Plot s versus r.

The accumulated proportion of variation (V) is calculated from the proportion of variation accounted for by each principal component (P in Step 3). Each principal component (PC_i) is weighted according to the proportion of total variance that it explains; as many components are derived as are necessary to account for the total variation in the data set. The proportion of variation of each principal component (V_i) is then normalized such that V_i is assigned 100% accumulated proportion, V_2 less than 100%, and so on to the point where 0%

is associated with no variation accounted for by any principal component, that is, scale of 0–100% on the abscissa of the plot of deviation in principal component score (Y_i) versus variability accounted for by principal components (V_i) drawn in Step 4. Sample classification is then carried out by plotting the slope (S) versus the correlation coefficient (r) computed from the Y versus V plot for each sample (Step 5). PCS is applicable to situations when unusual samples are discovered and/or when there are not enough data available to clearly classify these new samples into some of the known groups, and provided that there are sufficient principal components derived from the data.

6.2.5 Cluster analysis (CA)

CA attempts to determine structural characteristics of a data set by organizing the data into clusters or hierarchies based on distance or similarity between data points (variables, or individual samples or cases). The resulting structure is displayed as a dendrogram, which demonstrates the extent of similarity among each pair of data points.

6.3 Application of PCS

In this chapter, each of the above techniques and their various combinations will be examined using nonfood- and food-related data sets. All computations were performed using the SYSTAT package (Systat, 1988). A simple program was written for Steps 2 and 3 of PCS and the remaining PCS computations were carried out using SYSTAT and SYGRAPH (Systat, 1988). Also, Steps 2–5 of PCS were programmed in the Lotus 1-2-3 package (Que Corporation, 1989).

6.3.1 Nonfood application

Owing to the novelty of the PCS technique, it has yet to be applied in food research. Therefore, in order to first present, then compare, advantages and disadvantages of the new PCS technique with other methods of MVA, data used by Manly (1986) were employed (Table 6.1). These data demonstrate the employment population distribution of different European countries. Table 6.2 shows principal component scores for each country. Only four principal components (PC) are shown, with eigenvalues of 3.487, 2.130, 1.099 and 0.995 for PC1 to PC4, respectively. These values are all larger or equal to 1.0 (after rounding off) and much larger than the eigenvalue of 0.543 for PC5. Figure 6.1 shows the plot of PC1 versus PC2: while Eastern European countries were readily grouped, European Community (EC) and nonEC Western countries were intermingled. The result of LDA for classifying the countries belonging to EC and nonEC Western countries and Eastern European countries is shown in Figure 6.2. Except for the misclassification of Sweden into the EC group,

Table 6.1   Percentages of people employed in nine different industry groups in Europe (AGR = agriculture; MIN = mining; MAN = manufacturing; PS = power supplies; CON = construction; SER = service industries; FIN = finance; SPS = social and personal services; TC = transport and communications)[a]

Country	AGR	MIN	MAN	PS	CON	SER	FIN	SPS	TC
Belgium	3.3	0.9	27.6	0.9	8.2	19.1	6.2	26.6	7.2
Denmark	9.2	0.1	21.8	0.6	8.3	14.6	6.5	32.2	7.1
France	10.8	0.8	27.5	0.9	8.9	16.8	6.0	22.6	5.7
W. Germany	6.7	1.3	35.8	0.9	7.3	14.4	5.0	22.3	6.1
Ireland	23.2	1.0	20.7	1.3	7.5	16.8	2.8	20.8	6.1
Italy	15.9	0.6	27.6	0.5	10.0	18.1	1.6	20.1	5.7
Luxembourg	7.7	3.1	30.8	0.8	9.2	18.5	4.6	19.2	6.2
Netherlands	6.3	0.1	22.5	1.0	9.9	18.0	6.8	28.5	6.8
UK	2.7	1.4	30.2	1.4	6.9	16.9	5.7	28.3	6.4
Austria	12.7	1.1	30.2	1.4	9.0	16.8	4.9	16.8	7.0
Finland	13.0	0.4	25.9	1.3	7.4	14.7	5.5	24.3	7.6
Greece	41.4	0.6	17.6	0.6	8.1	11.5	2.4	11.0	6.7
Norway	9.0	0.5	22.4	0.8	8.6	16.9	4.7	27.6	9.4
Portugal	27.8	0.3	24.5	0.6	8.4	13.3	2.7	16.7	5.7
Spain	22.9	0.8	28.5	0.7	11.5	9.7	8.5	11.8	5.5
Sweden	6.1	0.4	25.9	0.8	7.2	14.4	6.0	32.4	6.8
Switzerland	7.7	0.2	37.8	0.8	9.5	17.5	5.3	15.4	5.7
Turkey	66.8	0.7	7.9	0.1	2.8	5.2	1.1	11.9	3.2
Bulgaria	23.6	1.9	32.3	0.6	7.9	8.0	0.7	18.2	6.7
Czechoslovakia	16.5	2.9	35.5	1.2	8.7	9.2	0.9	17.9	7.0
E. Germany	4.2	2.9	41.2	1.3	7.6	11.2	1.2	22.1	8.4
Hungary	21.7	3.1	29.6	1.9	8.2	9.4	0.9	17.2	8.0
Poland	31.1	2.5	25.7	0.9	8.4	7.5	0.9	16.1	6.9
Romania	34.7	2.1	30.1	0.6	8.7	5.9	1.3	11.7	5.0
USSR	23.7	1.4	25.8	0.6	9.2	6.1	0.5	23.6	9.3
Yugoslavia	48.7	1.5	16.8	1.1	4.9	6.4	11.3	5.3	4.0

[a]Adapted from Manly, 1986.

Table 6.2   Factor scores for 26 European countries

| | FACTOR | | | |
	1 Agriculture and lack of service industries	*2* Mining and power supplies	*3* Financial and service industries and lack of mining	*4* Lack of industrialization
Belgium	−0.93	−0.04	0.86	−0.08
Denmark	−1.30	−1.09	0.59	0.44
France	0.02	−0.20	0.98	−0.43
W. Germany	−0.04	0.45	0.45	−0.32
Ireland	−0.32	0.37	0.35	0.82
Italy	0.08	−1.40	−0.07	−1.19
Luxembourg	0.37	0.59	0.18	−1.05
Netherlands	−0.90	−0.59	1.17	−0.24
UK	−0.85	1.23	0.95	0.59
Austria	0.06	0.83	0.68	−0.45
Finland	−0.92	0.47	0.62	0.73
Greece	0.56	−1.12	−0.56	0.42
Norway	−1.77	−0.67	−0.09	0.31
Portugal	0.40	−1.11	−0.07	−0.17
Spain	1.67	−0.64	0.93	−1.67
Sweden	−1.29	−0.38	0.61	0.67
Switzerland	0.68	−0.39	0.98	−1.62
Turkey	1.29	−1.57	−0.85	3.00
Bulgaria	0.26	−0.25	−1.39	−0.34
Czechoslovakia	0.30	1.18	−1.19	−0.63
E. Germany	−0.61	1.70	−1.19	−0.44
Hungary	−0.12	2.37	−1.07	0.42
Poland	0.42	0.26	−1.41	0.06
Romania	1.55	−0.30	−1.11	−0.67
USSR	−0.99	−0.87	−2.06	−0.06
Yugoslavia	2.35	1.17	1.70	1.91

the classification was almost complete. A plot of slope (s) versus correlation coefficient (r) determined from LDA, using former West Germany as a reference, is shown in Figure 6.3. EC countries and Eastern European countries were separated; however, nonEC Western countries were scattered. Turkey was located far away from other European countries owing to the deviation of PC4 (lack of industrialization) from that of West Germany (the reference country) (Figure 6.4). Although classification into three groups was incomplete, the scatter graph shown in Figure 6.3 was interesting, as it appeared that the distribution was probably also affected by attributes other than economy. Although the employment population may have been most strongly affected by political and economic factors, other effects such as geography, climate, culture and tradition could not be totally ignored. PCA illustrated by the PC2 versus PC1 plot (Figure 6.1) would appear to be less informative than the results of PCS shown in Figure 6.3, although the separation of Eastern European countries from Western European countries was distinct.

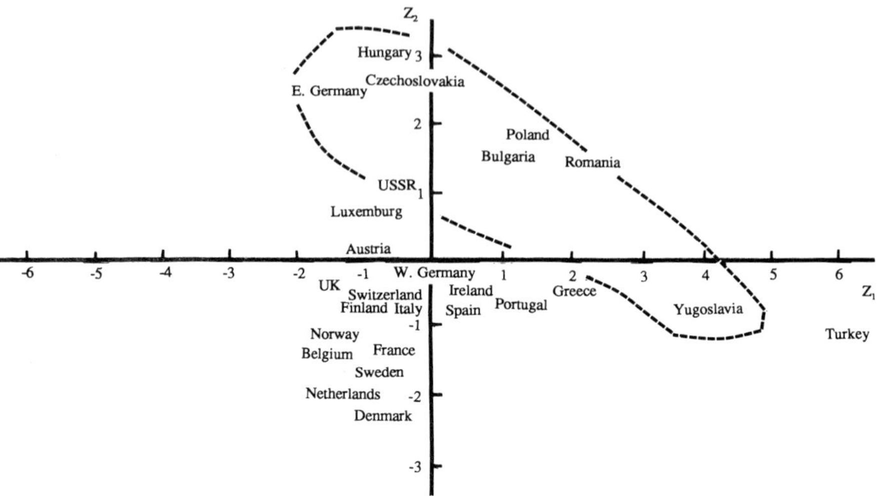

Figure 6.1 European countries plotted against the first two principal components, Z_1 and Z_2 for employment variables (adapted from Manly, 1986).

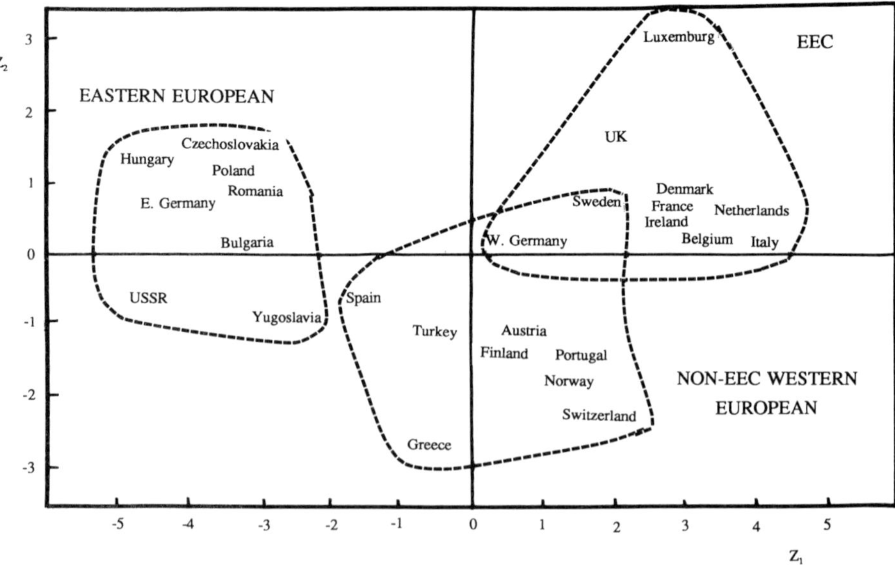

Figure 6.2 Plot of 26 European countries against their values for two canonical discriminant functions (adapted from Manly, 1986).

It is worth noting that many other plots can be generated using PCS by assigning other countries as the reference. For example, when former East Germany was used as the reference country, the PCS plot (Figure 6.5) demonstrated a different pattern from that when West Germany was used as the reference (Figure 6.3). In Figure 6.5 Eastern European countries could be seen in a block;

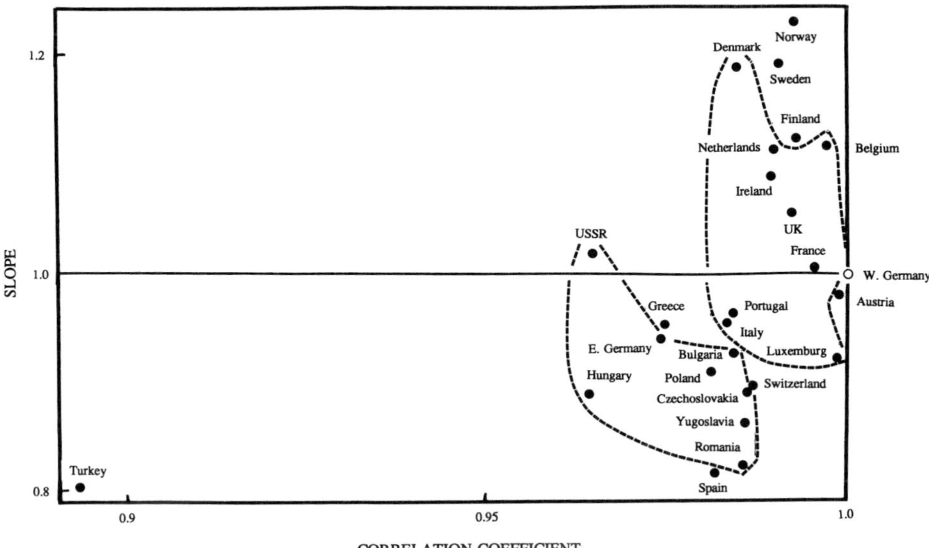

Figure 6.3 European countries plotted as slope versus correlation coefficient derived from principal component—similarity analysis using former West Germany as the reference.

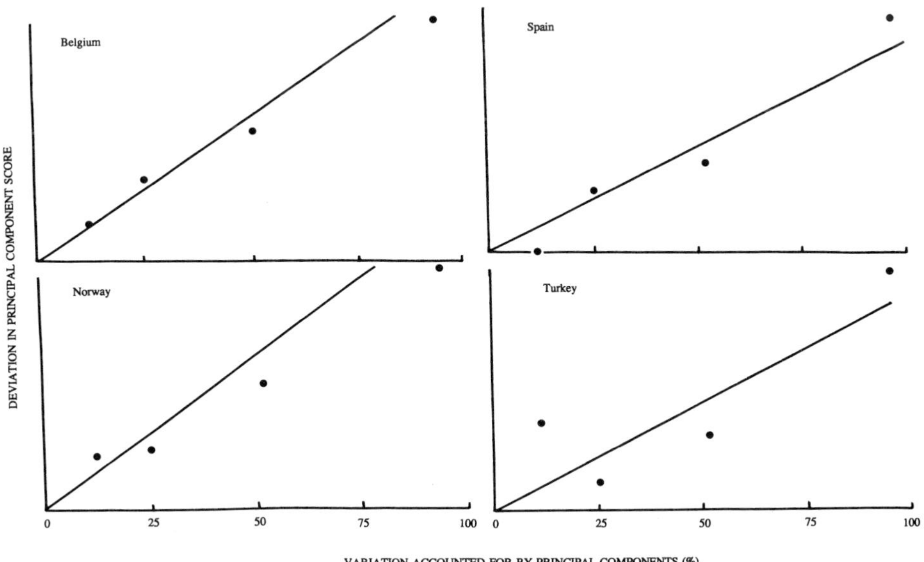

Figure 6.4 Principal component score plot of European countries. Deviations from the reference country (former West Germany) are plotted as differences from the reference line. The data points are first, second, third, and fourth principal components from right to left.

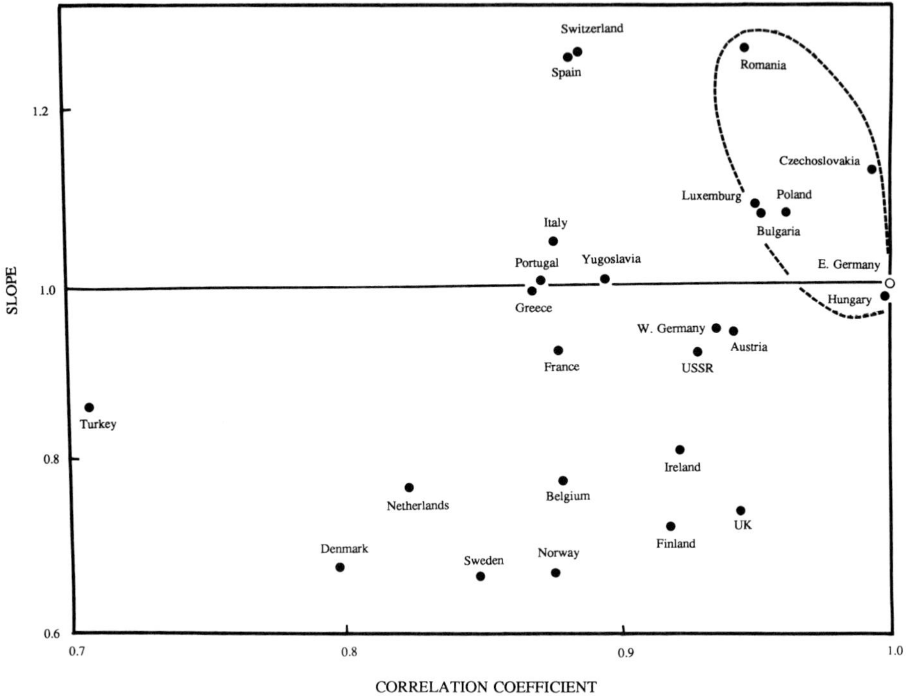

Figure 6.5 European countries plotted for slope versus correlation coefficient derived from principal component similarity analysis using East Germany as the reference.

however, the former USSR and Luxembourg were incorrectly classified. Distinction of EC from nonEC Western countries was not clear.

Figure 6.6 is the dendrogram shown in Manly's monograph (1986) generated from CA. Comparison of the classification shown in the dendrogram with that appearing in the PCS plot (Figure 6.3) indicates some dissimilarities. Although the classification of Eastern European countries was reasonable on the dendrogram, the UK, Italy and Luxembourg were misclassified. Another method of CA (positioning CA) revealed two different classifications (Figure 6.7). While the first cluster included most Western European countries, cluster 2 seemed to be mixed in the first classification. The six clusters shown in the second classification appeared reasonable except that cluster 4 included both Eastern and nonEC Western European countries. As Manly (1986) has stated, there is no generally acceptable 'best' method of CA; unfortunately, CA using different algorithms does not necessarily produce the same results on a given set of data. This is also the case in PCS, except that it has the advantage of providing more easily visualized clusters when compared with CA.

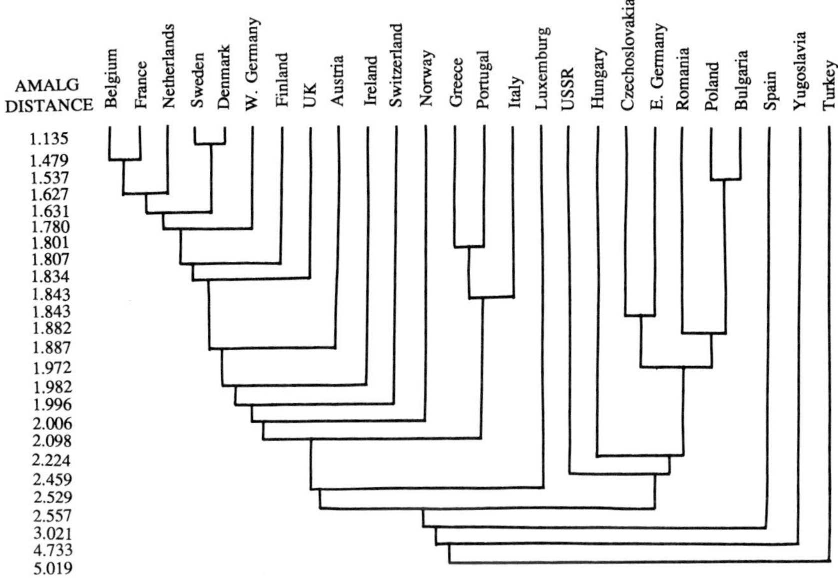

Figure 6.6 Dendrogram obtained from a nearest neighbour, hierarchical cluster analysis of data on employment in European countries.

(1)	(2)
Belgium	Greece
Denmark	Spain
France	Turkey
West Germany	Bulgaria
Ireland	Czechoslovakia
Italy	East Germany
Luxemburg	Hungary
Netherlands	Poland
UK	Romania
Austria	USSR
Finland	Yugoslavia
Norway	Portugal
Sweden	
Switzerland	

(1)	(2)	(3)	(4)	(5)	(6)
Luxemburg	East Germany	Turkey	Spain	Denmark	France
	Hungary	Yugoslavia	Bulgaria	Netherlands	West
	Czechoslovakia		Poland	UK	Germany
			Romania	Finland	Ireland
			USSR	Norway	Italy
			Portugal	Sweden	Austria
			Greece	Belgium	Switzerland

Figure 6.7 Classification results from positioning the cluster analysis in Figure 6.6.

6.3.2 Food-related applications

Chymosin is a milk-clotting enzyme with weak-proteolytic activity, despite a homology in the primary structure with pepsin (which possesses much stronger proteolytic activity). The diagonal plot method of Benyon (1982) to find matching sequences (Table 6.3) indicated that approximately 60% of the sequences were identical between the two enzymes (Yada, 1984). This homology was shown to be the highest among the different aspartyl proteinases. Physical/chemical properties of aspartyl proteinases (Table 6.4) were analyzed using PCA and yielded four principal components having eigenvalues greater than 1.0 (Table 6.5). The first principal component (PC1) was primarily concerned with CD data; PC2 included molar ellipticity at 193, 198, 200 and 202 nm, which were characteristic of α-helix, β-sheet, unordered, and β-turn, respectively; and PC3 and PC4 were both related to hydrophobicity, with the inclusion of charge effects in PC3. Two-dimensional plots involving the first three principal components are presented in Figure 6.8. In Figure 6.8, chymosin at pH 5.0, 5.3 and 8.0, as well as pepsin at pH 7.0 and 8.0, were not associated with their own groups, which may have been due to alterations in secondary structure with pH changes, reflected in the CD properties. In Figure 6.9, chymosin at a pH similar to that encountered during cheese-making (C3, C4, and C5) is more closely associated with the microbial proteinases than in Figure 6.8. However, this does not adequately explain the high milk-clotting/proteolysis ratios since *A. saitoi* proteinases and penicillopepsins (which exhibit low milk-clotting/proteolysis ratios) were also in close proximity. As shown in Figure 6.9, PC3 was capable of discriminating pepsin from chymosin on the basis of charge and hydrophobicity, which was consistent with the principal component loadings (Table 6.5).

Based on the assumption that parameters pertinent to charge and hydrophobicity of the same enzymes vary only slightly upon pH change, and thus contribute marginally to classification, CA was carried out (Aishima *et al.*, 1987). As illustrated in Figure 6.10, five distinct clusters were observed: three active enzyme groups (marked with letters H and M) G_1, G_3 and G_5 and two inactive

Table 6.3 Milk clotting/proteolysis ratio and similarity in sequences of aspartyl proteinases

Proteinases	Clot/proteolysis	Similarity[a]			
Chymosin	96				
Pepsin	47	57	25		
M. miehei	71	26			
M. pusillus	63			25	
					27
E. parasitica	60				
A. saitoi	0.6				
Penicillopepsin	0.8				

[a]Computed by the diagonal plot method

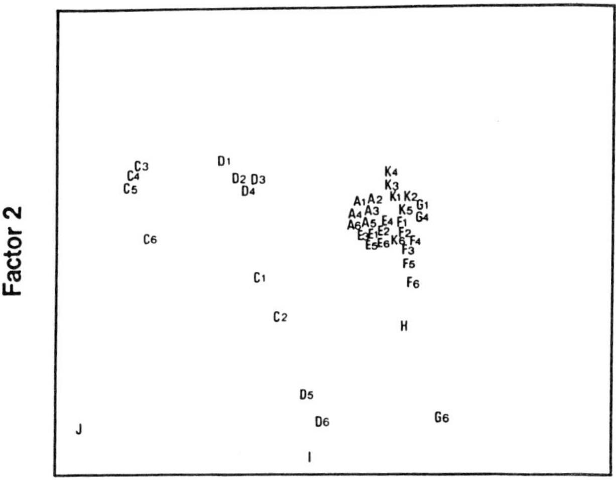

Figure 6.8 The plot of factor 2 versus factor 1 obtained from the principal component analysis of the various structural and intrinsic properties of the proteinases: A = *Mucor miehei* proteinase; K = *Endothia parasitica* proteinase; C = chymosin; D = pepsin; E = *Mucor pusillus* proteinase; F = *Aspergillus saitoi* proteinase; G = penicillopepsin; H = trypsin; I = α-chymotrypsin; J = papain The numbers 1, 2, 3, 4, 5, 6 represent pH 5.0, 5.3, 5.8, 6.3, 7.0, 8.0, respectively (adapted from Yada, 1984).

enzyme groups G_2 and G_4 (marked with a minus sign), based on milk-clotting activity. Enzymes included in all other independent subclusters were inactive. This result suggested the importance of secondary structure in milk-clotting activity. PCA was applied to 11 predictor variables yielding three factors with eigenvalues greater than 1.0. PCR, using these principal component scores, showed a significant correlation between milk-clotting activity and PC2, which was closely related to CD wavelengths of 198, 202 and 200 nm representing β-sheet, β-turn and random structures, respectively. The fractions of secondary structure, in addition to the above 11 predictors variables, were therefore used in LDA to discriminate aspartyl proteinases with high, medium and low milk-clotting activities. As shown in Figure 6.11, discrimination was successful but with some overlap among the groups. Overlap in grouping on canonical (i.e. discriminant) plots is not unusual for classification of continuous data such as milk-clotting activity. Based on the predictor variables/properties included in the discriminant functions, it was concluded that β-sheet and β-turn were the most important factors determining milk-clotting activity.

PCS analysis was applied to the data in Table 6.5 to visualize the effects of four principal components simultaneously, using chymosin as a reference. As shown in Figure 6.12, three bacterial rennets, i.e. aspartyl proteinase preparations of varied pH (A, B and E), were grouped together, thus demonstrating stability of

Table 6.4 Physical chemical measurements of several proteolytic enzymes[a]

Protein	pH	$[\theta]_{MRW\lambda}$ ($\times 10^{-3}$), deg cm^2 dmol^{-1}						
		190	193	198	200	202	210	213
M. miehei	5.0	97	−63	182	−159	−1040	−2711	−2787
proteinase	5.3	232	408	140	−443	−999	−2589	−2574
	5.8	428	177	489	−248	−1120	−2690	−2730
	6.3	313	−6	62	−356	−1286	−2811	−2820
	7.0	−762	−21	−126	−602	−1320	−3003	−2962
	8.0	643	−51	−364	−498	−1261	−2717	−2725
E. parasitica	5.0	1400	1402	542	86	−734	−2363	−2308
proteinase	5.3	866	1198	339	−31	−636	−2163	−2236
	5.8	1563	1303	683	−204	−731	−2203	−2423
	6.3	1141	1414	385	26	−836	−2492	−2377
	7.0	192	383	−4	−496	−938	−2237	−2250
	8.0	48	436	−697	−1486	−1889	−2537	−2492
Chymosin	5.0	1667	1337	−1449	−3373	−5133	−7438	−7015
	5.3	1978	1604	−3359	−4897	−5745	−6434	−5798
	5.8	7320	8960	2692	−906	−3971	−9761	−9391
	6.3	8912	8913	2788	−1125	−4610	−10460	−9577
	7.0	9650	8655	2454	−1308	−4169	−9680	−9399
	8.0	8171	6488	406	−3104	−5853	−10048	−9476
Pepsin	5.0	4726	5028	2591	494	1935	−7425	−7357
	5.3	3414	4514	2423	29	−2149	−7182	−7178
	5.8	3392	4174	1963	21	−2265	−7014	−6702
	6.3	2684	4094	2051	−47	−2398	−7144	−7059
	7.0	−436	−186	−5199	−6414	−7530	−6770	−6042
	8.0	−1776	−2295	−5842	−7088	−7528	−6847	−6086
M. pusillus	5.0	19	−869	−366	−1014	−1842	−3030	−3060
proteinase	5.3	1228	−522	−879	−1323	−2029	−3000	−2918
	5.8	532	−237	−429	−1184	−1884	−3208	−3067
	6.3	17	−496	−680	−1091	−1888	−3028	−2948
	7.0	−66	−780	−786	−1172	−1965	−3189	−3044
	8.0	−65	−1226	−1369	−1791	−2111	−2990	−2820
A. saitoi	5.0	640	769	71	−779	−1696	−2146	−1996
proteinase	5.3	866	805	−457	−944	−1463	−2193	−2142
	5.8	154	258	−831	−1471	−2027	−2024	−2063
	6.3	145	247	−957	−1378	−1821	−2085	−2019
	7.0	15	−722	−1787	−2028	−2319	−2396	−2246
	8.0	−503	−1267	−2316	−2737	−2742	−2182	−2082
Penicillopepsin	5.0	−266	842	519	−231	−796	−2212	−1959
	6.3	−494	214	238	−704	−1352	−2419	−2114
	8.0	−3201	−6430	−6467	−6195	−5208	−3388	−3042
Trypsin	6.3	−1793	−3050	−3185	−3365	−3560	−2945	−2568
Chymotrypsin	6.3	4110	870	−6204	−8830	−9730	−7270	−6200
Papain	6.3	14820	7590	−6990	−8880	−8165	−12580	−11590

[a]Source: Adapted from Yada and Nakai, 1986.
[b]Wavelengths in nanometers.
[c]ZP = ζ potential.
[d]H_ϕ(av) = Bigelow average hydrophobicity.
[e]CHG = ratio of acidic to basic amino acid residues.
[f]ASA = accessible surface area.
[g]ANS = apparent surface hydrophobicity using 1-anilino-8-napthalenesulfonate.
[h]CPA = apparent surface hydrophobicity using *cis*-parinaric acid.
[i]MC/PA = milk-clotting to proteolytic activity ratio.

222	224	225[b]	ZP[c]	$H_\Phi(av)$[d]	CHG[e]	ASA[f]	ANS[g]	CPA[h]	MC/PA[i]
−1398	−1103	−953	−14.7	1109	2.0	13000	2.0	21.0	114.31
−1456	−958	−833	−19.2	1109	2.0	13000	2.0	21.0	101.96
−1418	−1035	−833	−26.0	1109	2.0	13000	2.0	21.0	97.58
−1575	−1168	−952	−32.9	1109	2.0	13000	2.0	21.0	71.05
−1736	−1206	−915	−38.4	1109	2.0	13000	2.0	21.0	6.93
−1622	−1263	−1020	−45.3	1109	2.0	13000	2.0	21.0	0.00
−1442	−1151	−1031	−9.3	923	2.4	12900	7.0	113.0	82.11
−1322	−1094	−920	−14.9	923	2.4	12900	7.0	113.0	62.52
−1542	−1272	−1079	−20.5	923	2.4	12900	7.0	113.0	62.49
−1462	−1193	−1050	−26.1	923	2.4	12900	7.0	113.0	60.36
−1343	−1061	−952	−30.1	923	2.4	12900	7.0	113.0	0.00
−2018	−1790	−1670	−33.9	923	2.4	12900	7.0	113.0	0.00
−5589	−4813	−4532	−6.9	1120	1.6	11300	48.0	96.0	159.61
−4610	−4209	−3961	−9.5	1120	1.6	11300	48.0	96.0	132.81
−8191	−7485	−7139	−11.7	1120	1.6	11300	48.0	96.0	97.06
−8393	−7952	−7656	−14.0	1120	1.6	11300	48.0	96.0	95.75
−8135	−7444	−7037	−17.4	1120	1.6	11300	48.0	96.0	13.64
−8115	−7100	−6793	−20.8	1120	1.6	11300	48.0	96.0	0.00
−5700	−4158	−3756	−20.2	1063	10.7	12300	1.0	6.0	74.43
−4768	−3892	−3517	−25.0	1063	10.7	12300	1.0	6.0	57.89
−4795	−3857	−3465	−29.5	1063	10.7	12300	1.0	6.0	53.12
−4953	−3893	−3402	−34.6	1063	10.7	12300	1.0	6.0	47.43
−4102	−3444	−3136	−41.0	1063	10.7	12300	1.0	6.0	0.00
−4038	−3511	−3255	−52.2	1063	10.7	12300	1.0	6.0	0.00
−1477	−1063	−820	−7.47	1041	3.6	30600	7.0	3.0	127.30
−1434	−942	−712	−9.91	1041	3.6	30600	7.0	3.0	90.01
−1441	−922	−714	−13.91	1041	3.6	30600	7.0	3.0	86.04
−1532	−990	−742	−16.92	1041	3.6	30600	7.0	3.0	63.37
−1498	−1019	−824	−20.72	1041	3.6	30600	7.0	3.0	8.32
−1848	−944	−790	−26.91	1041	3.6	30600	7.0	3.0	0.00
−1077	−826	−711	−2.34	973	3.8	12200	6.0	73.0	1.45
−1115	−819	−794	−3.21	973	3.8	12200	6.0	73.0	0.93
−1337	−1083	−966	−4.17	973	3.8	12200	6.0	73.0	0.58
−1244	−972	−821	−5.79	973	3.8	12200	6.0	73.0	0.00
−1465	−1191	−1090	−8.65	973	3.8	12200	6.0	73.0	0.00
−1460	−1242	−1148	−13.90	973	3.8	12200	6.0	73.0	0.00
−261	132	252	−32.84	933	3.4	11600	3.0	23.0	22.95
−295	97	211	−39.46	933	3.4	11600	3.0	23.0	0.75
−1794	−1594	−1500	−47.56	933	3.4	11600	3.0	23.0	0.00
−1412	−1279	−1170	15.77	1034	1.9	9600	12.0	6.0	0.02
−4260	−4110	−4110	16.27	1030	0.7	9800	5.0	9.0	3.07
−11700	−11700	−11640	19.44	1159	0.6	8700	12.0	19.0	2.17

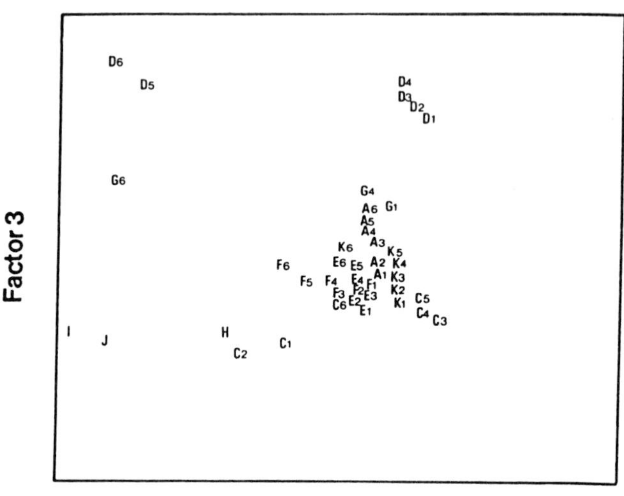

Figure 6.9 The plot of factor 3 versus factor 2 obtained from the principal component analysis of the various structural and intrinsic properties of the proteinases: A = *Mucor miehei* proteinase; K = *Endothia parasitica* proteinase; C = chymosin; D = pepsin; E = *Mucor pusillus* proteinase; F = *Aspergillus saitoi* proteinase; G = penicillopepsin; H = trypsin; I = α-chymotrypsin; J = papain. The numbers 1, 2, 3, 4, 5, 6 represent pH 5.0, 5.3, 5.8, 6.3, 7.0, 8.0, respectively (adapted from Yada, 1984).

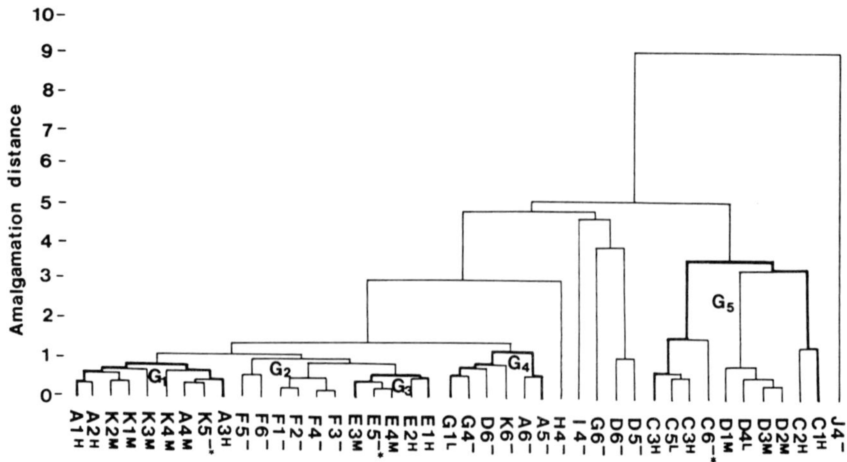

Figure 6.10 Dendrogram of 42 samples based on CD spectra and zeta potential: H = high activity; M = medium activity; – = no or low activity.

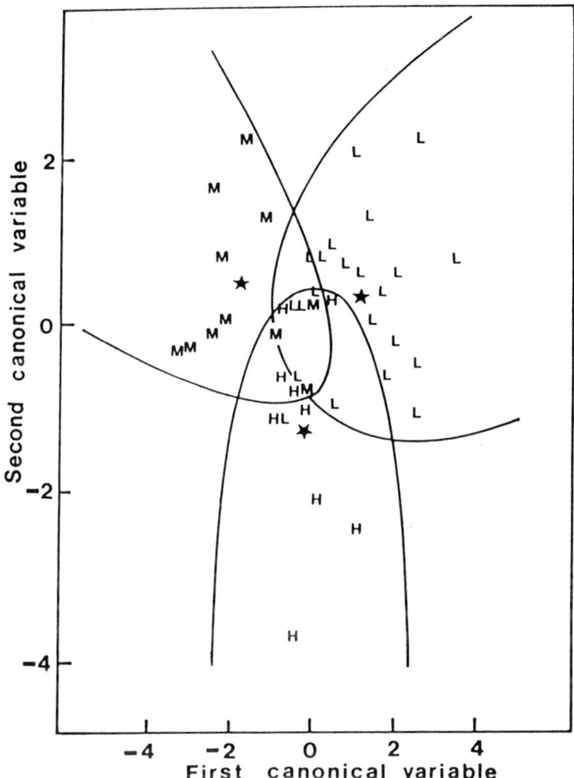

Figure 6.11 Canonical plot of 42 samples: H, M and L are samples of high, medium and low milk-clotting activity, respectively; ★ is the group mean.

these enzymes to pH changes. This grouping contrasts with the *M. miehei* proteinase located near acidic chymosin (Ca). While chymosin shifts towards acidic regions (C, Cb, Ca), pepsin (Da, D, Db) and penicillopepsin (Ga, G, Gb) show greater shift towards alkaline regions (Figure 6.12); *A. saitoi* proteinase showed a similar trend to these latter enzymes (i.e. Fa, F, Fb) but to a lesser extent. Enzymes inactivated in alkali (Eb, Cb and Ab), however, could not be differentiated. As shown in Figure 6.13, the greatest deviation of PC scores between pepsin and chymosin occurred in PC3; therefore, charge and hydrophobicity were two major parameters that distinguished these two enzymes (Table 6.5). This was similar to the conclusion drawn from the PC3 versus PC2 plot generated by PCA (Figure 6.9). This PCS plot shown in Figure 6.12 therefore appeared to be more informative than dendograms generated by CA (Figure 6.10).

The importance of β-sheets and β-turns in the mechanism of milk-clotting activity was demonstrated in the above study. It was postulated that charge and hydrophobicity were probably essential in the differentiation of pepsin and chymosin. The secondary structures of chymosin and pepsin predicted by the

Table 6.5   Factor loadings for the factors whose eigenvalues exceed 1.0[a]

Original variable	Factor loadings			
	Factor 1	Factor 2	Factor 3	Factor 4
213 nm	0.978	0.000	0.000	0.000
202 nm	0.974	0.000	0.000	0.000
210 nm	0.965	0.000	0.000	0.000
224 nm	0.954	0.000	0.000	0.000
225 nm	0.945	0.000	0.000	0.000
190 nm	−0.905	0.000	0.000	0.000
193 nm	−0.879	0.359	0.000	0.000
$H\Phi_{AVG}$	−0.719	0.000	0.000	0.394
ANS	−0.626	0.000	−0.495	−0.259
198 nm	0.000	0.971	0.000	0.000
200 nm	0.000	0.967	0.000	0.000
202 nm	0.537	0.811	0.000	0.000
Acidic/basic	0.000	0.000	0.870	0.000
Zeta potential	0.000	0.000	−0.692	0.000
ASA	0.000	0.000	0.000	0.794
CPA	0.000	0.000	−0.470	−0.780

[a]Source: adapted from Yada, 1984.

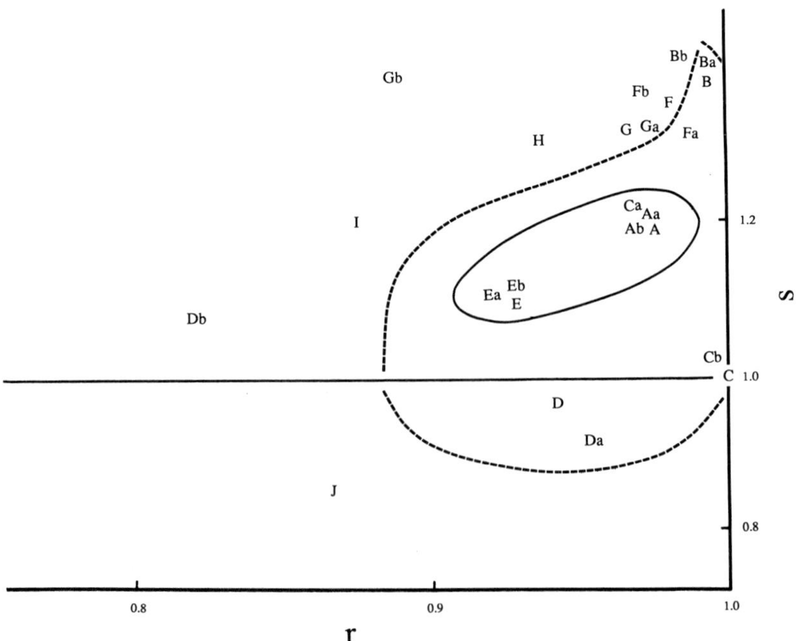

Figure 6.12 Plot (s versus r) of aspartyl proteinases with chymosin as a reference.

method of Chou and Fasman (1978) are illustrated in Figure 6.14. A marked difference was observed between sites 175–205, near the active site, with three β-turns in pepsin but not in chymosin. This portion of the sequence was located inside the crevice commonly formed between the active-site aspartic acid residues, that is, Asp_{33} (or Asp_{32} in pepsin) and Asp_{216} (or Asp_{215} in pepsin). Elimination of the three β-turns may be facilitated by replacing any one amino-acid residue of the four residues required for formation of a β-turn, for example, through off site-directed mutagenesis. As shown in Figure 6.15, changes in charge and hydrophobicity inside the crevices that permit accommodation of the para-κ-casein moiety of the substrate κ-casein may contribute to the characteristic enzymatic properties of chymosin that do not exist in pepsin.

Assuming that this hypothetical structural characterization explaining the enzyme-substrate relationship is tenable, it is not hard to understand the difficulty in explaining the full mechanism of milk-clotting from current data alone. More structure–function information is required concerning the active site, where enzyme and substrate interact, instead of structural data concerning the entire enzyme molecule. It is likely that at higher pH values, the crevices are wider and that the fit of substrate into the crevices is relaxed, resulting in a loss of milk-clotting activity. This change may not have, however, been detected by the techniques used in our previous work (Yada and Nakai, 1986).

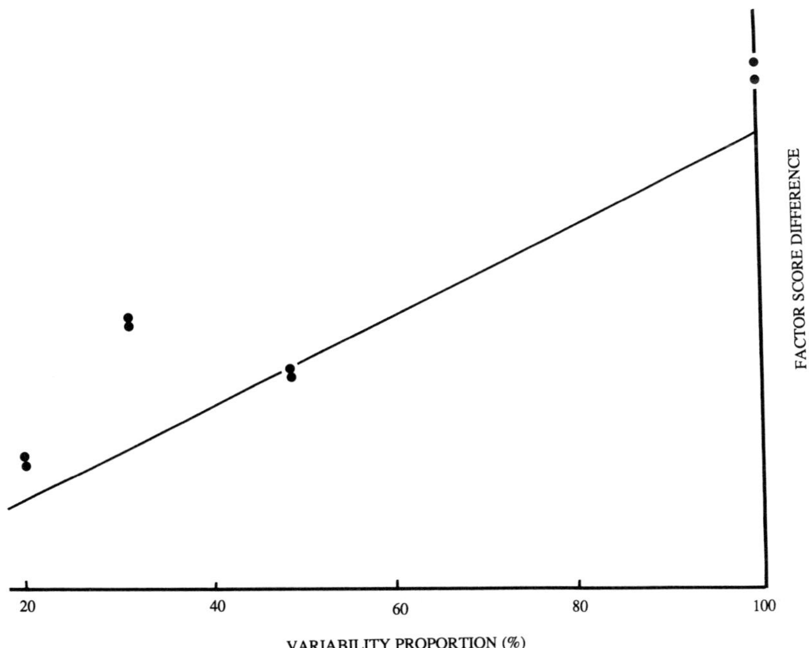

Figure 6.13 Factor score difference of pepsin versus chymosin.

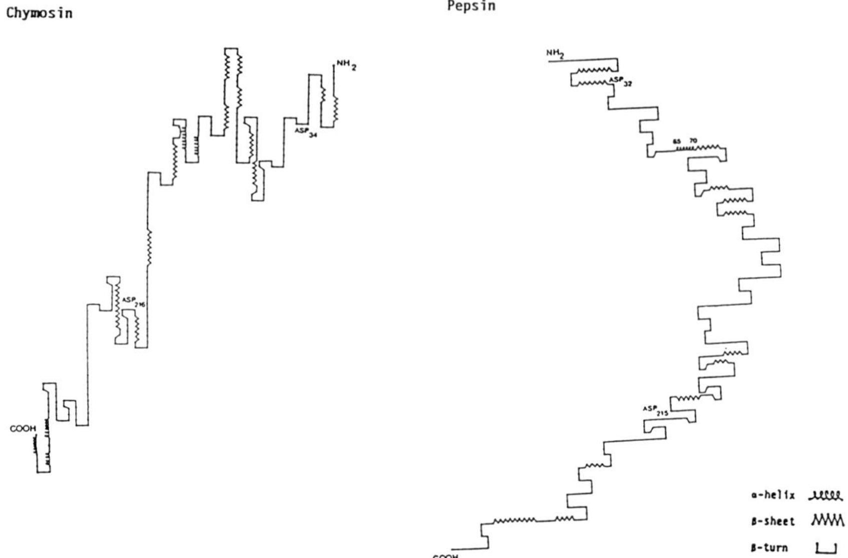

Figure 6.14 Comparison of secondary structures predicted by the method of Chou and Fasman (1978). (Adapted from Yada, 1984.)

Based on comparison of the results of MVA, it can be concluded that PCA and CA are useful in selecting effective predictors for classification, while PCS is useful in the classification of samples when the purpose of classification is not defined *a priori*. For supervised analysis such as discrimination of milk-clotting activity, PCR and LDA are useful, while CA can assist the selection of important predictors if the objective is discrimination.

6.4 Comparison of multivariate analyses

There are two major techniques for classifying samples using MVA, that is, classification into groups defined *a priori* (supervised) and classification using the original data to the greatest extent based on different principles (unsupervised). Since LDA is a supervised method, it allows for screening of data useful only for the given classification. Therefore, this technique is recommended in situations where there is a clearly defined, objectively selected classification, for example, differentiation of animal species or plant varieties or of a variety of food products prepared by different processes such as cheeses or wines. Since LDA uses portions of data useful only for classification purposes, if not properly used, important information may be lost or hidden in the original data. If the grouping of samples is not objectively made, such as the case when samples are grouped according to sensory test results, complete classification of samples

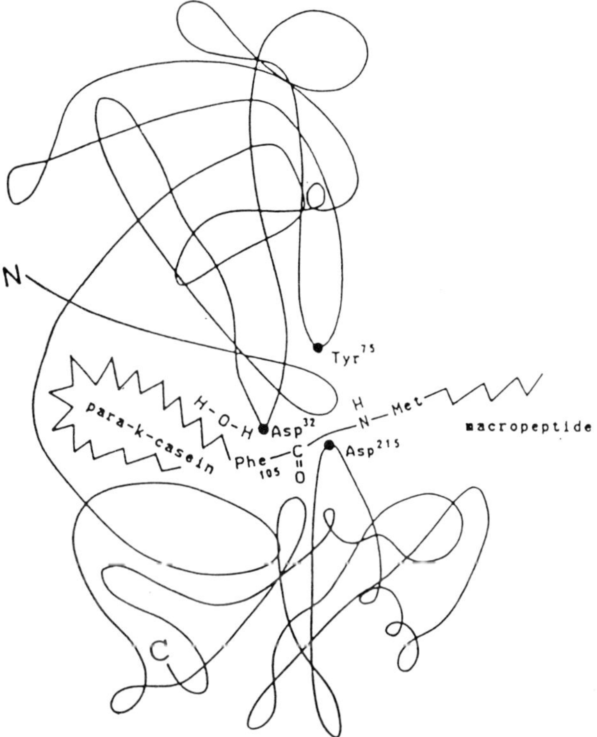

Figure 6.15 Hypothetical illustration of the action of chymosin on κ-casein.

may be difficult to achieve. For classification of continuous data, for example, protein denaturation, spoilage or functionality, the application of PCR is recommended since prediction or classification of unknown samples can be made more readily and accurately with the inclusion of additional statistical information, for example, coefficient of determination and standard error of estimate. The prediction equation thus derived may be particularly useful in investigations of food protein functionality, since it may be used for subsequent optimization of blends or formulations.

CA is an unsupervised method that is used in an attempt to utilize the maximum information pertaining to data variation. As seen in the above example of aspartyl proteinases, CA can be used for the prescreening of parameters useful for improved classification efficiency in subsequent analyses, for example, LDA. Since PCS attempts to use the maximum amount of information regarding data variation, in a manner similar to CA, the data obtained by PCS computation could be analogous to the results derived from CA (Vodovotz *et al.*, 1993). However, PCS and CA may not necessarily yield the same classification results, as demonstrated in the above examples, since the principles of the two methods

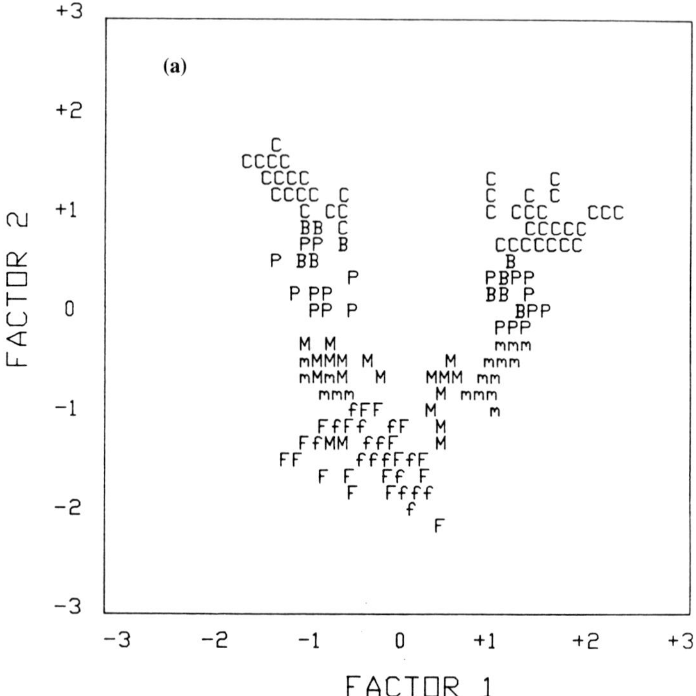

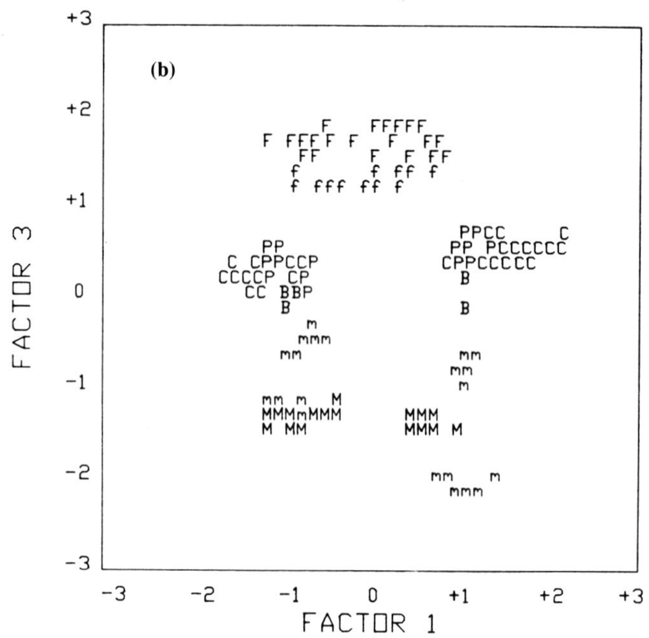

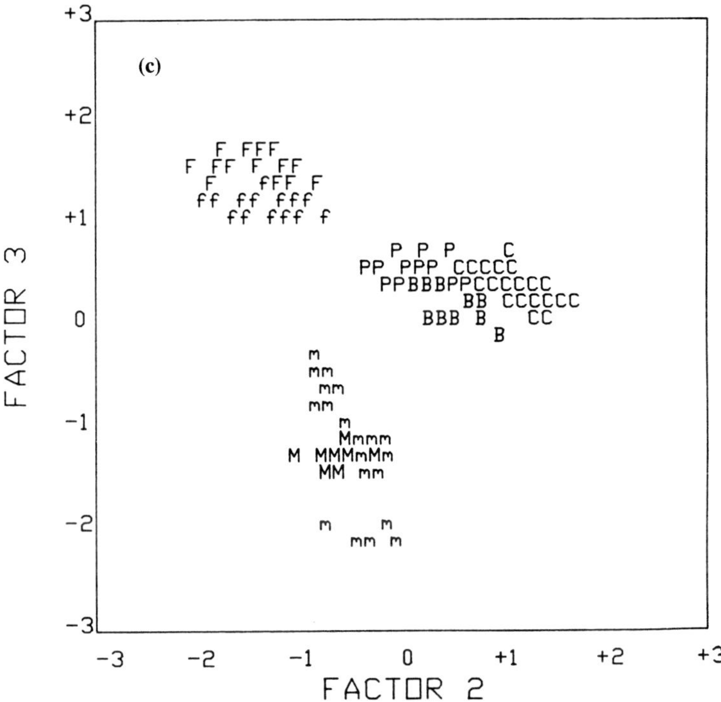

Figure 6.16 Plots of factor scores of unheated and heated meat samples obtained by principal component analysis of all variables excluding water-binding capacity: (a) factor 2 versus factor 1; (b) factor 3 versus factor 1; (c) factor 3 versus factor 2. Sample abbreviations: C, hand-deboned chicken; B, hand-deboned beef; P, hand-deboned pork; M, mechanically deboned pork; m, mechanically deboned chicken; F, cod fish; f, cod fish in presence of cryoprotectants. Overlapping factor scores are not shown. (From Li-Chan *et al.*, 1987, with permission.)

are quite different. It is our intention to use PCS as a new classification technique that can also be used for prescreening parameters for more efficient classification of samples. When data from an adequate number of samples have been accumulated, LDA should be carried out as the final, decisive classification measure for discriminating samples into clearly defined, objectively selected groups.

PCS may be useful when unusual samples are discovered and/or when there is not enough data available to classify clearly these new samples into some of the known groups. In cases where there may not be an adequate number of samples to use for LDA, PCS may be useful, either in finding similar groups (see Figures 6.3 and 6.5) or in locating new samples on a scatter graph among entire groups of samples. Characterization of Cheddar cheese samples ripened by accelerated methods is currently underway by applying MVA, in particular PCS, to HPLC data of water extracts obtained from samples. PCS may be useful in the classification of a new type of cheese in relation to known standard cheeses. The proposed technique is more useful when considerable portions of data variation are shared by many principal components. However, when only a few principal components account for most of the variation, such as in the case shown in Figure 6.16, PCS may not be superior to PCA. Further studies are required in which PCS is applied to food research problems before the method may be fully established. It is important, however, to realize that without using MVA technique for data processing, significant portions of information that could be derived from the data may be lost. The value of MVA techniques in food research, although not fully appreciated by many workers in the field, has been demonstrated.

References

Aishima, T. and Nakai, S. (1991) Chemometrics in flavor research. *Food Rev. Int.* **7**:33–101.

Aishima, T., Yada, R.Y., Mohler-Smith, A. and Nakai, S. (1987) Multivariate analysis of structure related data to explain milk-clotting activity of proteolytic enzyme. *J. Food Biochem.* **11**:121–32.

Benyon, R.J. (1982) The diagonal plot for sequence comparisons. *Biochem. Microcomputer Group* **7**:11.

Chou, P.Y. and Fasman, G.D. (1978) Empirical prediction of protein conformation. *Ann. Rev. Biochem.* **47**:251–76.

Girard, B. and Nakai, S. (1993) Species differentiation by multivariate analysis of headspace volatile patterns from canned Pacific Salmon, *J. Aquat. Food Prod. Technol.*, **2**:51–68.

Ennis, D.M., Beelens, H., Haring, H. and Bowman, P. (1982) Multivariate analysis in sensory evaluation. *Food Technol.* **36(11)**:83–90.

Jeon, I.J. (1991) Pattern recognition techniques for food research and quality assurance, in *Instrumental Methods for Quality Assurance in Foods* (eds D.Y.D. Fung and R.F. Matthews), Marcel Dekker, New York.

Li-Chan, E., Nakai, S., and Wood, D.F. (1987) Muscle protein–function relationships and discrimination of functionality by multivariate analysis. *J. Food Sci.* **52**:31–41.

Manly, B.F.J., (1986). *Multivariate Statistical Methods: A Primer*, Chapman and Hall, London.

Newell, G.J. and Lee, B. (1981) Ridge regression: an alternative to multiple linear regression for highly correlated data. *J. Food Sci.* **46**:968–71.

Que Corporation (1989) *Using Lotus 1-2-3: Release 3*. Que Corporation, Carmel, Indiana.

Systat (1988). *SYSTAT: The System for Statistics*. (ed. L. Wilkinson), SYSTAT Inc., Evanston, Illinois.

Vodovotz, Y., Arteaga, G.E., and Nakai, S. (1993) Principal component similarity analysis for classification and its application to GC data of mango. *Food Res. Int.*, **26**:355–63.

Yada, R.Y. (1984) *A Study of Secondary Structure Predictive Methods for Proteins and the Relationship Between Physical–chemical Properties and Enzymatic Activity of some Aspartyl Proteinases*. PhD Thesis, University of British Columbia.

Yada, R.Y. and Nakai, S. (1986). The use of principal component analysis to study the relationship between physical/chemical properties and milk-clotting to proteolytic activity ratio of some aspartyl proteinases. *J. Agric. Food. Chem.* **34**:675–79.

7 Surface adsorption of dairy proteins: Fouling of model surfaces

K.L. FULLER AND S.G. ROSCOE

Abstract

The surface adsorption behavior of the dairy protein β-lactoglobulin A at the open circuit platinum electrode has been investigated by dipping the electrode into a separate 0.16 g l^{-1} protein solution in a phosphate buffer (pH 7.0, 0.33 M ionic strength) thermostated to temperatures over the range 299–363 K. The charge density due to protein adsorption at the electrode surface was measured using cyclic voltammetry after transfer of the electrode to the electrochemical cell containing only the phosphate buffer. A well-defined dependence was obtained over the temperature range 299–363 K between the surface charge density for protein adsorption and the measured accessibility of free sulfhydryl groups, which is an indicator of the amount of conformational change, dissociation of dimeric units and/or aggregation of the protein. Protein removal was also investigated by measuring the current response during extended potential cycling in the phosphate buffer maintained at 273 K, 299 K and at the temperature of the dip solution. The results obtained suggest a possible treatment for the fouling of surfaces resulting from adsorption of β-lactoglobulin.

7.1 Introduction

Over the years the surface adsorption of macromolecules has drawn considerable interest because of the problems encountered by the food industry confronted with the 'fouling' of metal surfaces during processing (Hegg and Larsson, 1981; Lund and Sandu, 1981; Sivik and Hallström 1981; Almas and Lund, 1984; Skudder *et al.*, 1986; Kirtley and McGuire, 1989). The dairy protein, β-lactoglobulin, has been the subject of numerous publications because of its importance in the 'fouling' of metal surfaces and in the formation and stabilization of dairy foams and emulsions (Arnebrant and Nylander, 1986; Arnebrant *et al.*, 1987; Dickinson *et al.*, 1989; Kim and Lund, 1989). Experiments by Tissier *et al.* (1984) have shown that β-lactoglobulin is the dominant protein in the mechanism of deposit formation at the inner surface of a plate-sterilizer. A correlation was observed between the maximum denaturation rate of β-lactoglobulin and the maximum formation of deposits. A variety of techniques such as infra-red reflection–absorption spectroscopy (Liedberg *et al.*, 1986), ultraviolet (UV) spectroscopy (Cornell, 1984), circular dichroism (Cornell, 1982), small-angle X-ray scattering (Mackie *et al.*, 1991), ellipsometry (Arnebrant and Nylander, 1986; Arnebrant *et al.*, 1985, 1987), and measurement of radioactively labelled molecules at the metal surface (Rowley *et al.*, 1979) have been used in the investigation of the adsorption behavior of β-lactoglobulin. This study provides an electrochemical approach to investigating both the adsorption behavior of β-lactoglobulin and the conditions for optimization of removal of this protein from this model system. Previously, good agreement was observed for the surface concentrations obtained for adsorption of β-lactoglobulin at the platinum electrode surface (a model system) (Roscoe and Fuller, 1992, 1993) and stainless steel pellets (Kim and Lund, 1989) under similar experimental conditions. This suggests that the adsorption behavior of β-lactoglobulin is similar for the two metals.

This chapter describes research using the platinum electrode to investigate the surface adsorption behavior of the dairy protein β-lactoglobulin A using cyclic voltammetry. Anodic reactions in aqueous solution at the platinum electrode have been well characterized (Conway and Angerstein-Kozlowska, 1981) and processes involving the surface adsorption behavior of β-lactoglobulin have been thoroughly described elsewhere (Roscoe and Fuller, 1992, 1993). The technique employing the anodic underpotential deposition (upd) region to measure surface charge density provides a very sensitive technique for determining the adsorption behavior of the protein with the electrode surface. The results obtained previously (Roscoe and Fuller, 1992, 1993) suggest that β-lactoglobulin denatures at the electrode surface and, from the plateau regions observed, further denaturation or surface rearrangement occurs to accommodate the higher concentration of protein. The higher surface concentrations obtained with high-anodic end potentials of 1.0 V probably result from a potentially driven compression of the flattened proteins to allow greater interaction of carboxyl groups at the positively charged electrode surface. At the anodic potentials used, adsorption accompanied by electron transfer occurs through the carboxyl groups in a manner similar to that observed in the adsorption of amino acids (Marangoni *et al.*, 1989, 1991).

7.2 Experimental technique

7.2.1 Electrochemical cell and electrodes

Three compartment all glass cells were constructed to provide two glass-sleeved stopcocks as two of the separate compartments. Johnson Matthey and Mallory (Aesar, Toronto, Ontario) high-purity grade Pt wires were degreased by refluxing in acetone, sealed in soft glass, electrochemically cleaned by potential cycling in 1 M sulfuric acid (H_2SO_4) and stored in 98% H_2SO_4. The real surface area of the platinum electrode was obtained from the charge under the hydrogen upd peaks determined by cyclic voltammetry in 0.5 M H_2SO_4 at 298 K, taking 210 µc cm^{-2} as the charge required for the formation of a monolayer of H in the usual way (Angerstein-Kozlowska, 1984). The reference electrodes used were saturated calomel electrodes made according to a standard procedure (Kennedy, 1984). Their potentials were checked frequently against a standard hydrogen electrode and compared with the literature value (Kennedy, 1984). The saturated calomel electrodes were found to be reproducible to within ±1 mV.

7.2.2 Solutions

Phosphate buffer solutions at pH 7.0 and ionic strength 0.33 M were prepared using anhydrous monobasic potassium phosphate (KH_2PO_4) (cell culture tested; Sigma Chemical Company, St Louis, Missouri) and sodium hydroxide (prepared from the standardized Aldrich Acculute concentrate, Aldrich Chemical Company, Milwaukee, Wisconsin). Conductivity water (Nanopure water, resistivity = 18.3 megohm cm^{-1}), which had been distilled from acid potassium permanganate to remove organic impurities, was used to prepare the solutions. Solutions for calibration of the surface areas of the platinum electrode were prepared from Aristar grade 98% purity sulfuric acid (BDH Chemicals Limited, Dartmouth, Nova Scotia). β-Lactoglobulin A was prepared by ammonium sulfate precipitation (Armstrong et al., 1967) and purified by HPLC with a DEAE-5PW column (Bio-Rad, Mississauga, Ontario) with detection by UV adsorbance at 280 nm. A gradient flow of 1 ml min^{-1} of 100% buffer A (10 mM imidazole + 0.05 M NaCl at pH 7.0) to 100% buffer B (10 mM imidazole + 0.4 MNaCl at pH 7.0) was used over a period of 35 min. The sample was subsequently extensively dialysed against distilled water and freeze-dried.

7.2.3 Methods

Cyclic voltammograms were obtained using a potentiostat (model HA-301, Hokuto Denko, Axial Precision Instruments, Oakville, Ontario) and a function generator (model HB-111, Hokuto Denko, Axial Precision Instruments, Oakville, Ontario) to produce a repeating triangular potential function. The sweep rate used throughout was 500 mV s^{-1}. The potential and current were taken from the outputs of the potentiostat and measured by an X–Y recorder/plotter (model 720M, Allen Datagraph, QC Instruments, Mississauga, Ontario).

The method for characterization of the protein on the electrode surface and the experimental methods have been described in detail elsewhere (Roscoe and Fuller, 1992, 1993). In order to characterize the behavior of β-lactoglobulin A at the platinum electrode, it was necessary in each experiment to first determine the surface charge density of oxide deposition at the platinum electrode by cyclic voltammetry under the same experimental conditions as those used with the proteins. From the cyclic voltammograms the surface charge density (Q_{ADS}) resulting from adsorption of the protein was determined from the integrated current-potential response corresponding to anodic oxidation in the presence of protein (Q_{Oo}^{p}) and to oxide reduction (Q_{Or}^{p}). The difference calculated between the surface charge density for anodic oxidation and that for oxide reduction was attributed to surface adsorption or oxidation of species other than O or OH present in these aqueous solutions. A small difference in the integrated surface charge density from the cyclic voltammograms of the phosphate buffer calculated from the difference between anodic oxide (Q_{Oo}) and the oxide reduction charge (Q_{Or}) was subtracted from those determined for the protein solutions. The remaining charge density was attributed to protein surface adsorption:

$$Q_{ADS} = (Q_{Oo}^{p} - Q_{O}^{p}) - (Q_{Oo} - Q_{Or}) \tag{7.1}$$

The surface charge density due to protein adsorption accompanied by electron transfer was corrected for oxide formation in the absence of protein during these potentiodynamic sweeps in the buffer solutions. Corrections were also made for the charge due to the capacitance effect of the double layer. This technique uses the 'blocking' of the oxide deposition as a measure of the efficiency of the protein to adsorb at the electrode surface. Although this 'blocking' by the protein is not as extensive as that observed for glycine (Marangoni *et al.*, 1989) and α- and β-alanine (Marangoni *et al.*, 1991), the technique gives a sensitive response to the adsorption behavior of the protein.

This technique has been used in the present research to investigate the adsorption behavior of β-lactoglobulin A at the open circuit platinum electrode by dipping the electrode into a separate 0.16 g l^{-1} protein solution in a phosphate buffer (pH 7.0, 0.33 M ionic strength) thermostatted to temperatures over the range 299–363 K. The charge density due to protein adsorption at the electrode surface was measured using cyclic voltammetry after transfer of the electrode to the electrochemical cell containing only the phosphate buffer. Protein removal was also investigated by measuring the current response during extended potential cycling in the phosphate buffer maintained at 273 K, 299 K and at the temperature of the dip solution.

These results were correlated with the accessibility of the free sulfhydryl group determined under conditions of varying temperature using 5,5'-dithiobis(2-nitrobenzoate) or DTNB (Ellman's reagent) with spectroscopic measurements (Glazer *et al.*, 1975; Jantova *et al.*, 1968). This accessibility of the free sulfhydryl group is an indication of the amount of conformational change, dissociation of dimeric units and/or aggregation of the protein. The procedure was modified slightly in order to determine the free sulfhydryl groups under the present experimental conditions with phosphate buffer and in the absence of EDTA and denaturing solvents. The solutions were held at the experimental temperature for 30 min. DTNB was then added and allowed to react for an additional 30 min before absorbance measurements were made at 412 nm using a UV/vis spectrophotometer (model DU-65, Beckman Instruments, Mississauga,

Ontario). Calculations of concentrations were made using the value of $\epsilon_{412} =$ 13 600 M^{-1} cm^{-1} (Janatova *et al.*, 1968).

7.2.4 Results

7.2.4.1 Effect of protein adsorption on an open-circuit electrode.   In the previous studies the surface adsorption behavior was investigated during potentiodynamic sweeping of a platinum electrode to the anodic end potentials 0.4 and 1.0 V (Roscoe and Fuller, 1992, 1993). In the present study, however, the adsorption behavior of the protein at the open circuit electrode followed by a characterization of the surface charge density by potential sweeping was investigated. In order to study the extent of surface adsorption by the protein independent of the applied potential, a series of experiments were made in which the electrode was first characterized by potentiodynamic cycling in the phosphate buffer at pH 7.0. Following this, the electrode was removed from the electrochemical cell after stopping the cycle in the double-layer region at −0.1 V. The electrode was then dipped for a measured length of time in a separate solution of the same phosphate buffer maintained at the same temperature as the electrochemical cell. The electrode was then transferred back to the electrochemical cell, containing only phosphate buffer, which was pre-potentiostatted to immediately allow an anodic sweep from −0.1 V followed by cycling to characterize the oxide reduction area. This was repeated for a series of experiments in which the electrode was held for longer periods of time in the dip solution before being transferred to the electrochemical cell. The measured amount of oxide formed on the electrode appeared to reach a plateau value after about 2 min holding time in the dip solution.

The experiment was then repeated dipping the electrode for measured lengths of time in a solution of β-lactoglobulin A at a concentration of 0.16 g l^{-1} in the phosphate buffer at pH 7.0. The electrode adsorbates were determined in the same manner, after transferring the electrode back to the electrochemical cell, which contained only the phosphate buffer (Figure 7.1). The surface charge density due to protein adsorption, was corrected for oxide deposition from the buffer solution. The dipped electrode was not rinsed before transferring to the electrochemical cell to prevent removal of loosely bound protein in the outer layers. The results for adsorption at the open circuit electrode of β-lactoglobulin at 299 K showed a plateau level with increase in holding time in the protein-buffer solution (Figure 7.2). The charge due to surface adsorption on the open circuit electrode was similar within experimental error to that obtained previously with potential cycling in β-lactoglobulin solutions (Roscoe and Fuller, 1992, 1993). There was slightly more scatter in the results of these experiments due to repeated removal and repositioning of the electrode. Despite the latter, the results indicated that at this temperature β-lactoglobulin adsorbed readily at the metal surface independent of an applied potential.

7.2.4.2 Effect of temperature on protein adsorption. As the temperature was increased to that encountered during processing, the adsorption of β-lactoglobulin also increased. Figure 7.2 shows the results for a series of experiments in which surface charge density was measured by potential cycling at temperatures 299, 343, 348, 353, 358, and 363 K after dipping the open-circuit electrode in the protein solutions at the same temperatures for measured lengths of time.

At the higher temperatures, the increased surface adsorption probably resulted from an increase in the availability of the carboxyl groups due to conformational changes of the denatured protein. Measurements made on the accessibility of the sulfhydryl groups during temperature denaturation using Ellman's reagent have shown a steady increase in the measured sulfhydryl over the temperature range 299–343 K (Figure 7.3). However, at temperatures 348–363 K, a marked decrease in the accessibility of the sulfhydryl group was measured and probably reflected denaturation followed by agglomeration of β-lactoglobulin molecules over this temperature range (Sawyer, 1968; De Witt and Swinkels, 1980).

Experiments were also made in which the electrode was dipped in the β-lactoglobulin solution at 358 K and then measured in three separate experiments

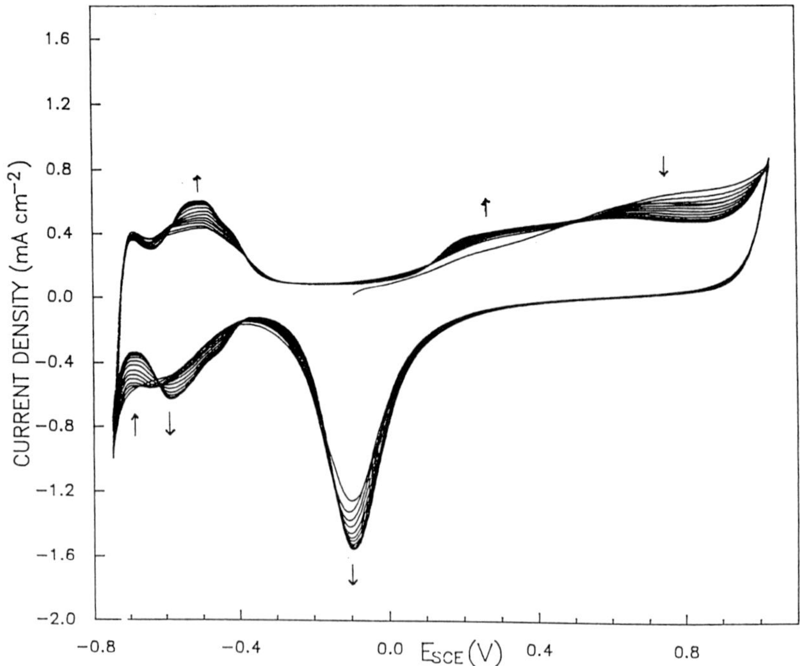

Figure 7.1 Cyclic voltammogram of the platinum electrode in 0.1 M phosphate buffer (pH 7.0, 299 K) after transferring the dipped electrode in a solution of 0.16 g l⁻¹ β-lactoglobulin A in phosphate buffer (pH 7.0, 299 K) for 16 min. Sweep rate 500 mV s⁻¹.

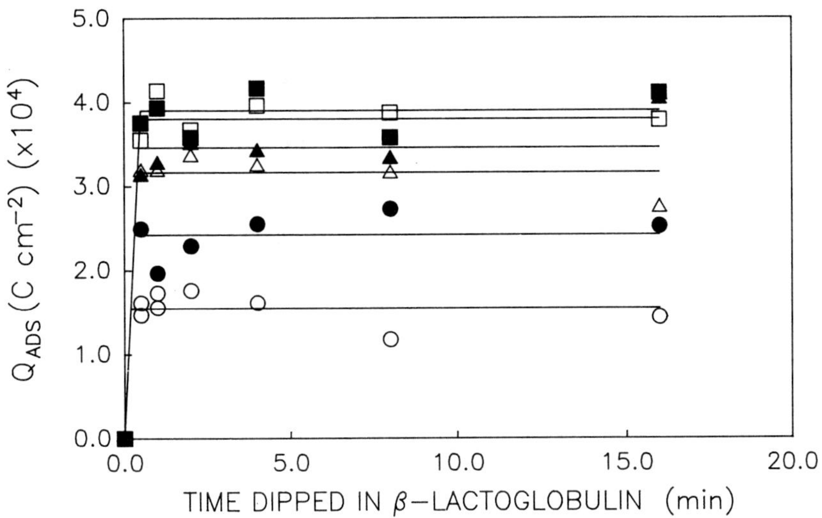

Figure 7.2 Surface charge density (Q_{ADS}) of β-lactoglobulin A at a platinum electrode versus the time the electrode was dipped in the 0.16 g l^{-1} β-lactoglobulin solution in phosphate buffer (pH 7.0) and measured at the same temperatures: ○, 299 K; ●, 343 K; △, 348 K; ▲, 353 K; □, 358 K; ■, 363 K.

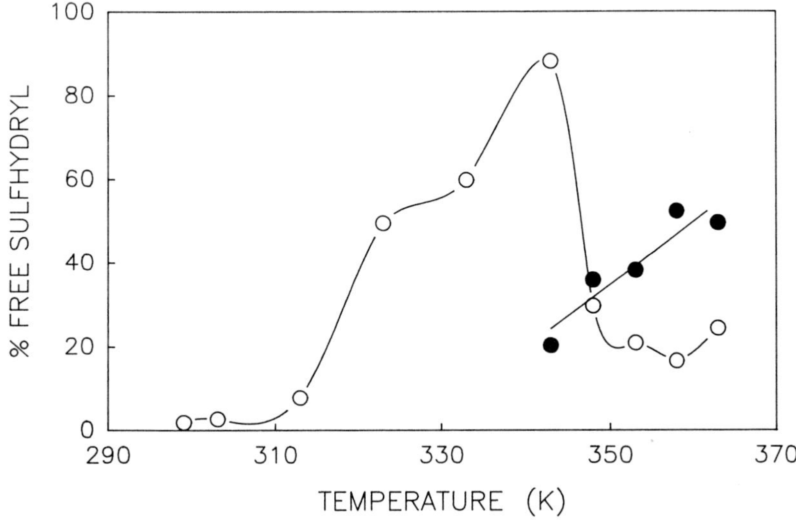

Figure 7.3 Percent free sulfhydryl versus temperature (K): ○, protein heated for 30 min at the specified temperature; ●, protein heated for 30 min at the specified temperature, then cooled to 299 K for 30 min.

in the electrochemical cell thermostatted at 358, 299, and 273 K (Figure 7.4). Although adsorption occurred at the same temperature of 358 K, the measured surface charge density was found to decrease with decreasing temperature. A similar behavior was obtained when the dip temperature was 343 K with measurements made at 343, 299, and 273 K (Figure 7.5).

When the dip temperature was 299, 343, and 358 K, and the temperature of the electrochemical cell maintained at 299 K, the measured surface charge density was similar for all three experiments (Figure 7.6). Measurements were also made of the accessibility of the free sulfhydryl of the protein at 299 K following denaturation at the higher temperatures. The results in Figure 7.3 showed that less agglomeration occurs accompanied by conformational change exposing the free sulfhydryl group. A linear response could be seen in the percentage of free sulfhydryl of the partially renatured protein measured after cooling to 299 K as a function of increasing temperature above the critical temperature for denaturation of 343 K. The results were consistent with those reported by De Witt and Swinkles (1980).

A similar behavior was observed when the dip temperatures were 343 and 358 K (Figures 7.4 and 7.5) and the electrochemical cell maintained at 273 K. The small surface charge density compared well with results obtained previously for β-lactoglobulin adsorption during potential cycling at 273 K since very little surface adsorption occurred at these temperatures (Roscoe and Fuller, 1992, 1993). Reliable results for the measurements of the accessible free sulfhydryl groups at 273 K could not be obtained due to moisture condensation on the cuvette during the spectroscopic measurements.

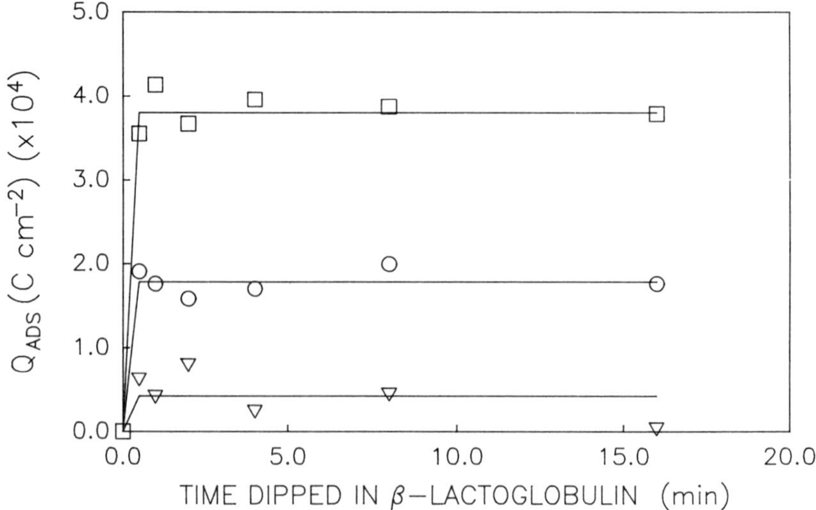

Figure 7.4 Surface charge density (Q_{ADS}) of β-lactoglobulin A at a platinum electrode versus the time the electrode was dipped in the 0.16 g l^{-1} β-lactoglobulin solution in phosphate buffer (pH 7.0) at 358 K and measured at the temperatures: ▽, 273 K; ○, 299 K; □, 358 K.

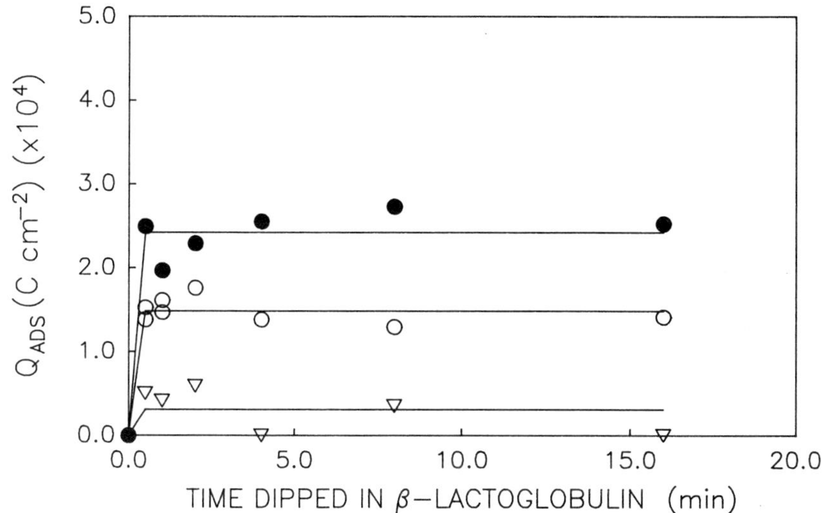

Figure 7.5 Surface charge density (Q_{ADS}) of β-lactoglobulin A at a platinum electrode versus the time the electrode was dipped in the 0.16 g l⁻¹ β-lactoglobulin solution in phosphate buffer (pH 7.0) at 343 K and measured at the temperatures: ▽, 273 K; ○, 299 K; ●, 343 K.

7.2.4.3 Effect of temperature on protein removal. The next step was to investigate the possible methods for removal of the protein from the metal electrode surface. Adsorption on the electrode surface was accomplished as in the above experiments by dipping the open circuit electrode for a measured length of time in a solution of 0.16 g l⁻¹ β-lactoglobulin in phosphate buffer (pH 7.0, μ = 0.33 M). This solution was maintained at a specified temperature in a separate water bath. Following this, the electrode was transferred to the electrochemical cell containing only the buffer solution, which was also maintained at a specific preset temperature. This was done so that the adsorption and removal processes could be investigated separately as a function of temperature. The removal of protein was measured by following potential cycling from –0.75–1.0 V for an extended period of time. A series of experiments was made in which the surface charge density for adsorption at 299, 343, 348, 353, 358, and 363 K was compared with that for removal of protein at 273 and 299 K, as well as at the temperature of adsorption. This was done to see if the protein could be removed by potential cycling at the same temperature at which adsorption occurred, or if protein removal would be more efficient by cooling the surface to room temperature or even to 273 K. The latter temperature was of interest as previous results had shown that very little surface adsorption resulted with β-lactoglobulin under potential cycling at 273 K (Roscoe and Fuller, 1992, 1993). Therefore protein removal should be favorable under these conditions.

After continuous potential cycling for several minutes, the profile of the cyclic

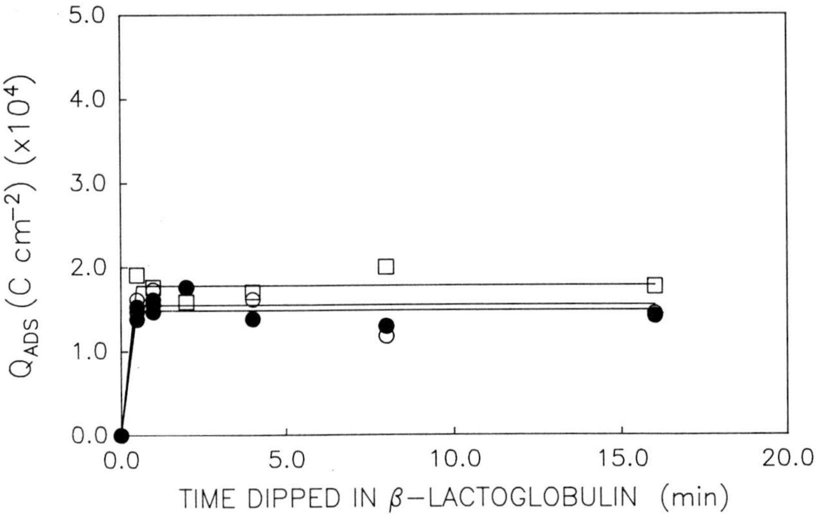

Figure 7.6 Surface charge density (Q_{ADS}) of β-lactoglobulin A at a platinum electrode versus the time the electrode was dipped in the 0.16 g l^{-1} β-lactoglobulin solution in phosphate buffer (pH 7.0) at the following temperatures and measured at 299 K: ○, 299 K; ●, 343 K; □, 358 K.

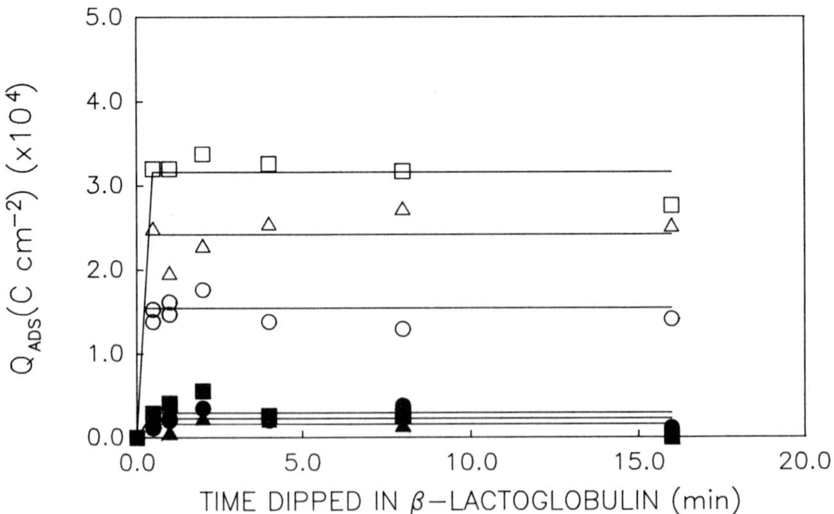

Figure 7.7 Surface charge density (Q_{ADS}) of β-lactoglobulin A at a platinum electrode versus the time the electrode was dipped in the 0.16 g l^{-1} β-lactoglobulin solution in phosphate buffer (pH 7.0) and measured at the same temperatures: ○, 299 K; △, 343 K; □, 348 K. The surface charge density (Q_{ADS}) after cycling for 2 min: ●, 299 K; ▲, 343 K; ■, 348 K.

voltammogram returned to that characteristic of the phosphate buffer system in the absence of the protein. The surface charge density due to protein adsorption was negligible following this extended cycling after correcting for oxide formation from the buffer solution in the usual way (Figures 7.7 to 7.9). These experiments suggest that, within the experimental error of the measurements, protein adsorbed during dipping without an applied potential can be measured using this technique. However, with continued cycling in the buffer solution, β-lactoglobulin was subsequently removed and did not remain permanently adsorbed onto the surface of the electrode.

7.2.4.4 Effect of sweeptime on protein removal. The protein removal process was then measured as a function of time by potential cycling from −0.75–1.0 V. This was carried out in the electrochemical cell following adsorption on the open circuit electrode held for 16 min in a solution of 0.16 g l^{-1} β-lactoglobulin in phosphate buffer (pH 7.0), which was maintained at a specified temperature in a water bath. The effect of sweeptime on protein removal was studied for a series of experiments in which protein adsorption in the dip solution and protein removal in the electrochemical cell containing only the phosphate buffer were carried out at the same temperature. Figure 7.10 shows the results obtained with measurements made at 299, 343, and 348 K. The times required for removal of the protein from the electrode surface under potential cycling were shorter at the higher temperatures. With a sweep rate of 500 mV s^{-1} over the potential range −0.75–1.0 V, protein removal was complete after approximately six cycles.

When the temperatures for both adsorption and protein removal were increased to 353, 358 and 363 K, removal occurred rapidly but incompletely (Figure 7.11). It was of interest, therefore, to investigate the adsorption of the protein at these higher temperatures followed by potential cycling at the lower temperatures of 299 and 273 K. The results for protein adsorption at 358 K, followed by potential cycling at 358, 299 and 273 K are shown in Figure 7.12. Although removal of protein was slower at the lower temperatures, complete removal was observed after 2 min sweeptime. Results from adsorption at 343, 358 and 363 K followed by potential cycling at 273 K also showed complete removal of the protein (Figure 7.13).

7.3 Discussion

This study has demonstrated that the surface interfacial behavior of β-lacto-globulin is very temperature-dependent and that surface adsorption increases with denaturation. In addition, surface adsorption of β-lactoglobulin occurred readily at the open circuit electrode with a resulting surface charge density, measured by potential cycling to the anodic potential 1.0 V, comparable with that obtained during adsorption at the anodically polarized electrode (Roscoe and Fuller, 1992, 1993). This method for investigating surface adsorption behavior has been shown

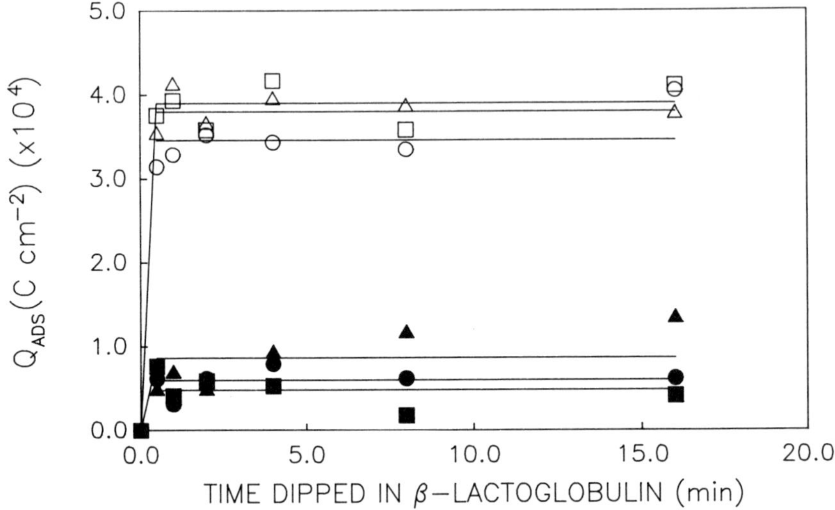

Figure 7.8 Surface charge density (Q_{ADS}) of β-lactoglobulin A at a platinum electrode versus the time the electrode was dipped in the 0.16 g l^{-1} β-lactoglobulin solution in phosphate buffer (pH 7.0) and measured at the same temperatures: ○, 353 K; △, 358 K; □, 363 K. The surface charge density (Q_{ADS}) after cycling for 2 min: ●, 353 K; ▲, 358 K; ■, 363 K.

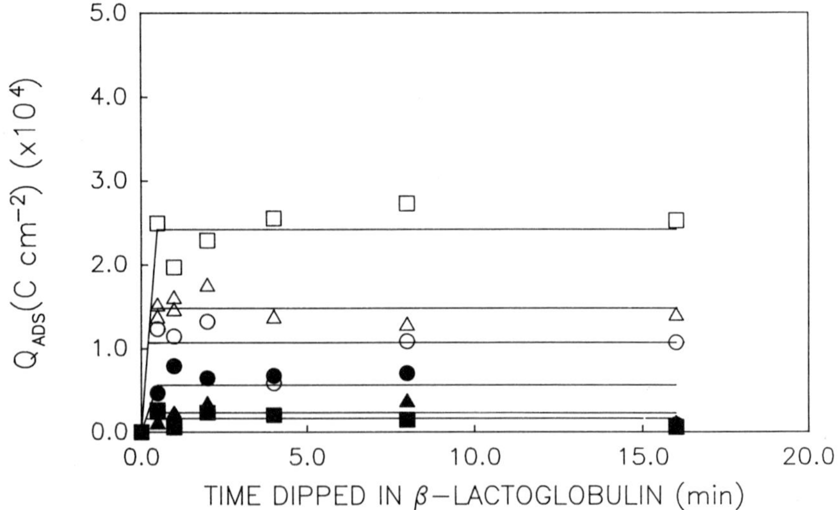

Figure 7.9 Surface charge density (Q_{ADS}) of β-lactoglobulin A at a platinum electrode versus the time the electrode was dipped in the 0.16 g l^{-1} β-lactoglobulin solution in phosphate buffer (pH 7.0) at 343 K and measured at the temperatures: ○, 273 K; △, 299 K; □, 343 K. The surface charge density (Q_{ADS}) after cycling for 2 min: ●, 273 K; ▲, 299 K; ■, 343 K.

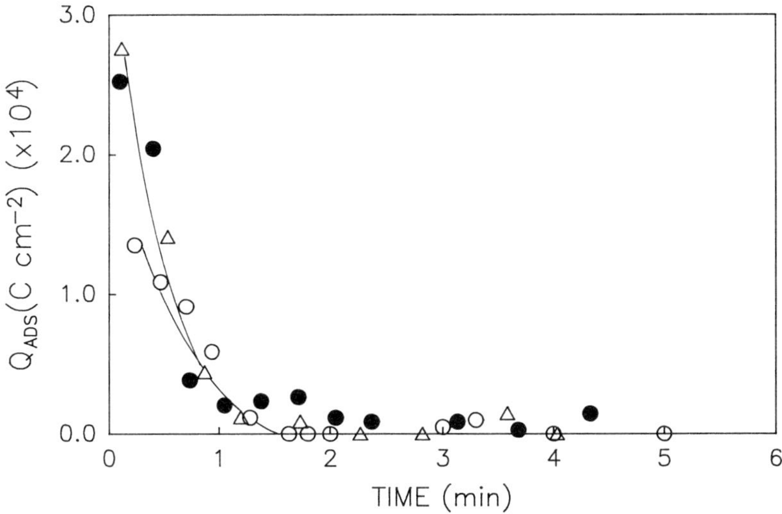

Figure 7.10 Surface charge density (Q_{ADS}) of β-lactoglobulin A at a platinum electrode versus the time of potential sweeping after the electrode was dipped in the 0.16 g l^{-1} β-lactoglobulin solution in phosphate buffer (pH 7.0) and measured at the same temperatures: ○, 299 K; ●, 343 K; △, 348 K.

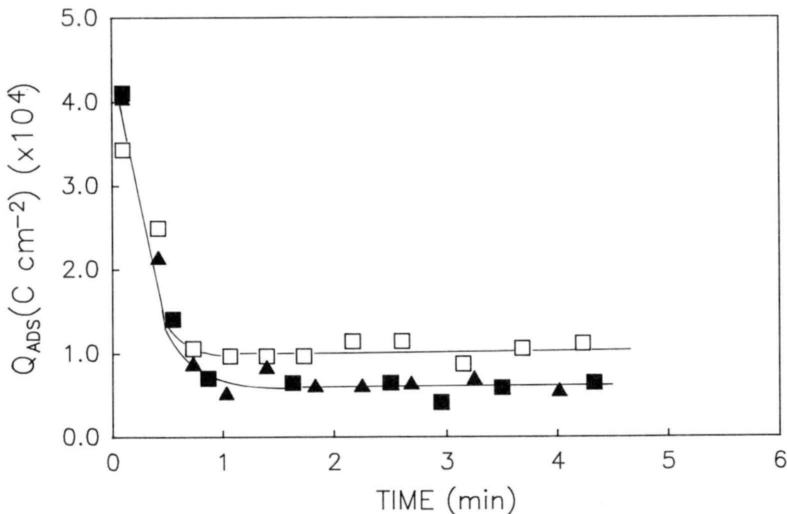

Figure 7.11 Surface charge density (Q_{ADS}) of β-lactoglobulin A at a platinum electrode versus the time of potential sweeping after the electrode was dipped in the 0.16 g l^{-1} β-lactoglobulin solution in phosphate buffer (pH 7.0) and measured at the same temperatures: ▲, 353 K; □, 358 K; ■, 363 K.

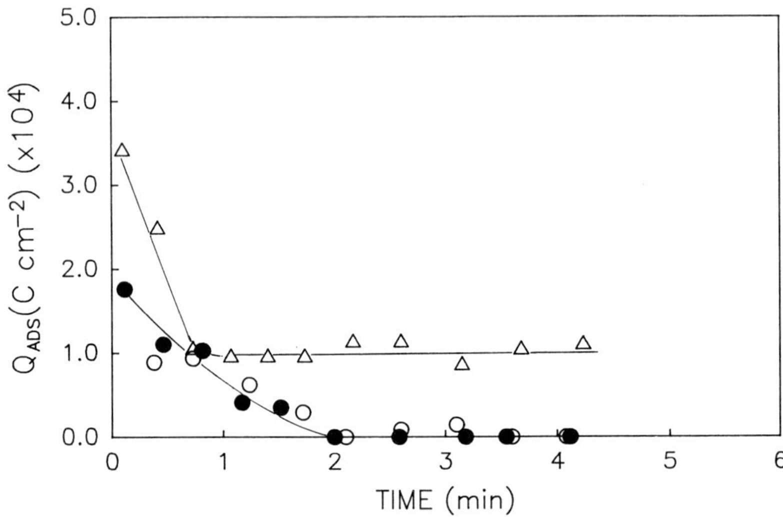

Figure 7.12 Surface charge density (Q_{ADS}) of β-lactoglobulin A at a platinum electrode versus the time of potential sweeping after the electrode was dipped in the 0.16 g l^{-1} β-lactoglobulin solution in phosphate buffer (pH 7.0) at 358 K and measured at the temperatures: O, 273 K; ●, 299 K; △, 358 K.

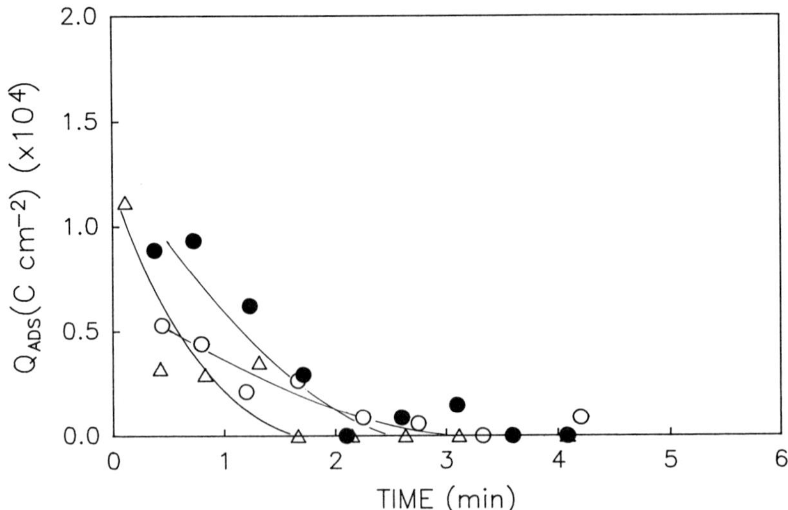

Figure 7.13 Surface charge density (Q_{ADS}) of β-lactoglobulin A at a platinum electrode versus the time of potential sweeping after the electrode was dipped in the 0.16 g l^{-1} β-lactoglobulin solution in phosphate buffer (pH 7.0) at the following temperatures and measured at 273 K: O, 343 K; ●, 358 K; △, 363 K.

to give good agreement with results obtained using other techniques such as small-angle X-ray scattering (Mackie *et al.*, 1991) and ellipsometry (Arnebrant and Nylander, 1986; Arnebrant *et al.*, 1987). The surface charge density results from the two-electron transfer process in the adsorption and oxidation of the carboxyl groups at the electrode surface of these anodic potentials (Roscoe and Fuller, 1992, 1993). The surface charge density resulting from adsorption at the open-circuit electrode reaches a plateau level even with very short dip times in the β-lactoglobulin solutions over the temperature range 299–363 K.

Most proteins show fairly rapid heat denaturation somewhere between 333–363 K, which affects the conformation of the proteins (Walstra and Jenness, 1984). Above 323 K, β-lactoglobulin dissociates into monomers, which is sometimes designated as a prenatured state. The previous results (Roscoe and Fuller, 1992, 1993) showed a change in the slope of ln (Q_{ADS}) versus T^{-1} at about 323 K, which suggested that denaturation occurred at this temperature. The present results showed a slightly different behavior because adsorption occurred at higher temperatures on an open-circuit electrode that was dipped into the protein solution. A gradual increase in the surface charge density occurred with increased temperature over the range 343–363 K.

At temperatures above 333 K an uncombined sulfhydryl in β-lactoglobulin is exposed by unfolding of the protein (Walstra and Jenness, 1984). Therefore, the sulfhydryl becomes available for reaction with various reagents and can be used as an index of the denaturation process. The results from the previous study (Roscoe and Fuller, 1992, 1993) showed a corresponding increase in the accessibility of the sulfhydryl group with increasing temperature. A change in slope of the plot of percentage free sulfhydryl versus temperature occurred at 323 K (Figure 7.3). Above this temperature β-lactoglobulin dissociates into monomers. A comparison of the surface charge density with the accessibility of the free sulfhydryl as measured spectroscopically over the temperature range 299–343 K, suggested that denaturation allows greater access and interaction of the protein's active sites with the electrode surface. Denaturation to give 10% exposed free sulfhydryl was sufficient to make a significant contribution to surface interaction as measured by surface charge density.

The present results have shown a decrease in the exposure of the free sulfhydryl group as the temperature is increased from 348 K to 363 K (Figure 7.3). This probably resulted from further conformational changes accompanied by agglomeration at these temperatures. In a differential scanning calorimetric study of the thermal denaturation of β-lactoglobulin in phosphate buffer at pH 6.8 at temperatures up to 373 K, De Witt and Swinkles (1980) reported that 343 K was a critical temperature for the denaturation of this protein. Wong (1989) found that when β-lactoglobulin at pH 7.0 is heated above 343 K, polymerization and aggregation began in the absence of other proteins. Aggregation results from increased sulfhydryl reactivity due to an increased intra- and intermolecular randomization of disulfide cross-links during heat treatments to 370 K (Sawyer, 1968). An increase of disulfide cross-links during heat-treatment of β-lactoglobulin A between 338 and

413 K has also been reported by Watanabe and Klostermeyer (1976). Thus heat-treatment may restrict complete unfolding near 353 K (Tanford, 1968; Creighton, 1977). Even when complications arising from disulfide exchange reactions are eliminated, evidence has been obtained that some residual structure remains after thermal treatment of β-lactoglobulin to 363 K (Ananthanarayanan *et al.*, 1977).

In an investigation of the structural changes involved in denaturation, Hegg and Larsson (1981) concluded that the denaturation temperature may be different for a protein molecule adsorbed at the interface compared with the same protein molecule in the bulk solution. Their results showed a drastic increase in film thickness of β-lactoglobulin at temperatures near the corresponding denaturation temperature. They found that two different processes influenced the film thickness, with one starting at about 323 K, which was proposed to be involved in fouling in milk exchangers, and the other at about 343 K. They also reported that the aggregation of β-lactoglobulin in aqueous solution started at temperatures considerably lower than the denaturation temperature, and that these two bulk changes were reflected in the film behavior. The present results also showed a change in the denaturation and adsorption behavior at the same two temperatures as noted above.

A reasonably gradual increase in surface adsorption was observed in the present and previous results (Roscoe and Fuller, 1992, 1993) compared with those obtained at the hydrophilic chromium surface where the adsorption increased only slightly from about 2.0 to 2.2 mg m^{-2} over the temperature range 298–346 K (Arnebrant *et al.*, 1987). When the temperature reached 348 K, the surface concentration increased abruptly to 8 mg m^{-2}. It was concluded that at these temperatures some conformational changes of the protein must occur before aggregation involving sulfhydryl groups begins at the surface.

Interesting results were obtained when protein removal was measured at a lower temperature than protein adsorption (Figure 7.9). The results suggested that protein adsorption on the open-circuit electrode at the higher temperatures, at which denaturation occurs in the bulk solution, probably included many layers, some of which were probably loosely bound. When the electrode was transferred to the electrochemical cell containing the buffer solution at a lower temperature, very fast thermal equilibration must have occurred with the metal electrode. A rapid response to the change in temperature appeared to be reflected by rapid conformational changes in the adsorbed protein to that characteristic of its adsorption behavior at that temperature. The surface charge density measurements reflected this structural reorientation (Figures 7.4–7.6).

Similar behaviour was observed previously (Roscoe and Fuller, 1993) for experiments in which the protein was denatured at 343 K in the bulk solution, subsequently allowed to cool to 299 K in the bulk solution, and then protein adsorption was measured by potential cycling in the solution containing the β-lactoglobulin in phosphate buffer. Under these conditions the renaturation of the protein was confirmed from the measurements of the accessibility of the free sulfhydryl groups. However, in the present research, the temperature-dependent structural changes of the protein occurred on the electrode surface.

This was important, as it suggested that protein adsorption was not accompanied by irreversible denaturation. Instead, renaturation may have occurred at the surface of the electrode as it did in solution with the exception that the oxidized protein underwent decarboxylation accompanied by formation of the next lower aldehyde at the interactive amino-acid residues (Marangoni *et al.*, 1989, 1991).

In a previous experiment (Roscoe and Fuller, 1993), the electrode was dipped in the β-lactoglobulin solution at 299 K, transferred to the electrochemical cell containing only the phosphate buffer at the same temperature and allowed to stand without an applied potential for 30 min. During this period the adsorbed protein remained on the electrode surface, as determined from the surface charge density measured subsequently by potential cycling. The protein was only removed from the electrode surface by continued cycling.

In the present investigation of the removal of β-lactoglobulin, potential cycling over the range −0.75–1.0 V resulted in complete oxidation and removal of protein following protein adsorption at 299–348 K. However, adsorption at 353–363 K gave only incomplete oxidation and removal. Instead, a lower plateau value was attained, which remained even with extensive potential cycling. It was interesting to note that these plateau levels ranged from approximately 1.8–2.8 mg m^{-2}, which corresponds to a monolayer surface coverage of geometrically flat β-lactoglobulin and to the surface concentration normally observed at room temperature by a variety of techniques (Mackie *et al.*, 1991). This plateau may depict a more stable configuration of the adsorbed protein on the electrode surface, which correlated well with the plateau value obtained during protein adsorption with potential cycling to 0.4 V (Roscoe and Fuller, 1992, 1993). The plateau value also corresponded to that observed for adsorption of β-lactoglobulin on stainless steel pellets (Kim and Lund, 1989). Complete oxidation and removal of protein was accomplished by decreasing the temperature to 299 and 273 K.

In order to consider why complete removal of the protein was attained after adsorption at temperatures between 299 and 348 K, but not at temperatures greater than 348 K (i.e., 353–363 K) with protein removal the same temperature, it was necessary to examine the surface coverage by protein at these high temperatures. The surface charge density reached a value of 400 μc cm^{-2} for the plateau values at 363 K (Figure 7.2). This corresponded to a surface coverage of θ = 2 and a surface concentration of Γ = 14.6 mg m^{-2}. This value would represent about four layers of geometrically undenatured β-lactoglobulin molecules. However, since it is more reasonable to expect denaturation to occur at the surface of the electrode, especially since denaturation has already been accomplished in the bulk solution at these temperatures, the observed results more likely represented many layers of denatured protein. Therefore, at these higher temperatures, it was conceivable that an extensive surface coverage developed such that complete removal was not achieved at the higher temperatures. Thus, within the double-layer region a continuous oxidation and reduction process resulted in protein interaction and removal accompanied by movement and conformational changes on the electrode surface. This caused sufficient protein to remain within the double-layer region of the electrode surface to give an effective monolayer surface coverage. When the temperature for protein removal by potential cycling was decreased to 299 K and 273 K, the conformational behavior of the molecule was altered substantially to give a much reduced surface interaction. At these lower temperatures, renatured protein probably moved away from the cold electrode surface and was lost in the bulk solution of the phosphate buffer. Previous results have shown very little adsorption of β-lactoglobulin at 273 K at the anodically polarized electrode (Roscoe and Fuller, 1992, 1993).

The above results have shown that with the present model, both adsorption and removal of protein are temperature-dependent due to the conformational behavior of the protein, both in the bulk solution and at the metal interface. By using this conformational behavior of β-lactoglobulin, accompanied by potential cycling, which oxidizes the carboxyl groups, protein removal was accomplished following adsorption of the protein over the temperature range 299–363 K from a phosphate buffer at pH 7.0. This model system has provided a different perspective on the surface adsorption behavior of β-lactoglobulin, with results that compare well with those reported in the literature using other techniques.

Acknowledgements

The authors wish to thank Dr Gilles Robitaille of the Department of Animal Science, Macdonald College of McGill University, Ste Anne de Bellevue, Quebec for preparing and purifying β-lactoglobulin A for these experiments. Grateful acknowledgement is also made to the Natural Sciences and Engineering Research Council, Canada, for support of this research.

References

Almas, K.A. and Lund, D.B. (1984) Cleaning and characterization of stainless steel exposed to milk. *Surface Technol.* **23**:29–39.

Ananthanarayanan, V.S., Ahmad, F. and Bigelow, C.C. (1977) The denaturation of β-lactoglobulin-A at pH 2. *Biochim. Biophys. Acta* **492**:194–203.

Angerstein-Kozlowska, H. (1984) Surfaces, cells, and solutions for kinetic studies, in *Comprehensive Treatise of Electrochemistry*, Vol. 9, (eds E. Yeager *et al.*), Plenum Press, New York, pp. 15–59.

Armstrong, J.McD., McKenzie, H.A. and Sawyer, W.H. (1967) On the fractionation of β-lactoglobulin and α-lactalbumin. *Biochim. Biophys. Acta.* **147**:60–72.

Arnebrant, T. and Nylander, T. (1986). Sequential and competitive adsorption of β-lactoglobulin and κ-casein on metal surfaces. *J. Coll. Interface Sci.* **111**:529–33.

Arnebrant, T., Barton, K. and Nylander, T. (1987) Adsorption of α-lactalbumin and β-lactoglobulin on metal surfaces versus temperature. *J. Coll. Interfac Sci.* **119**:383–90.

Conway, B.E. and Angerstein-Kozlowska, H. (1981). The electrochemical study of multiple-state adsorption in monolayers. *Acc. Chem. Res.* **41**:49–56.

Cornell, D.G. (1982) Lipid–protein interactions in monolayers: egg yolk phosphatidic acid and β-lactoglobulin. *J. Coll. Interfac. Sci.* **88**:536–45.

Cornell, D.G. (1984) Ultraviolet spectroscopy of β-lactoglobulin films. *J. Coll. Interfac. Sci.* **98**:283–85.

Creighton, T.E. (1977) Conformational restrictions on the pathway of folding and unfolding of the pancreatic trypsin inhibitor. *J. Mol. Biol.* **113**:275–94.

De Witt, J.N. and Swinkels, G.A.M., (1980) A differential scanning calorimetric study of the thermal denaturation of bovine β-lactoglobulin; thermal behaviour at temperatures up to 100°C. *Biochim. Biophys. Acta* **624**:40–50.

Dickinson, E., Mauffret, A., Rolfe, S.E. and Woskett, C.M. (1989) Adsorption at interfaces in dairy systems. *J. Soc. Dairy Technol.* **42**:18–22.

Glazer, A.N., De Lange, R.J. and Sigman, D.S. (1975) Chemical Modification of proteins, selected methods and analytical procedures, in *Laboratory Techniques in Biochemistry and Molecular Biology*, Vol. 4, Part I. (eds. T.S. Work and E. Work), North-Holland Publishing Company, Amsterdam, pp. 113–14.

Hegg, P.O. and Larsson, K. (1981) Ellipsometry studies of adsorbed lipids and milk proteins on metal surfaces, in *Fundamentals and Applications of Surface Phenomena Associated with Fouling and Cleaning in Food Processing*, (eds B. Hallström, D.B. Lund and Ch. Trägardh), pp. 250–55.

Janatova, J., Fuller, J.K. and Hunter, M.J. (1968) The heterogeneity of bovine albumin with respect to sulfhydryl and dimer content. *J. Biol. Chem.* **243**:3612–22.

Kennedy, J.H. (1984) *Analytical Chemistry*, Harcourt Brace Jovanovich, San Diego, pp. 481–84.

Kim, J.C. and Lund, D.B. (1989) Adsorption of β-lactoglobulin onto stainless steel surfaces, in *Fouling and Cleansing in Food Processing*, (eds. H.G. Kessler and D.B. Lund), Druckerei Walch, Augsburg, pp. 187–99.

Kirtley, S.A. and McGuire, J. (1989) On the differences in surface constitution of dairy product contact materials. *J. Dairy Sci.* **72**:1748–53.

Liedberg, B., Ivarson, B., Hegg, P.O. and Lundstrom, I. (1986) On the adsorption of β-lactoglobulin on hydrophilic gold surfaces: studies by infrared reflection-absorption spectroscopy and ellipsometry. *J. Coll. Interfac. Sci.* **114**:386–97.

Lund, D. and Sandu, C. (1981) State-of-the-art of fouling: heat transfer surfaces, in *Fundamentals and Application of Surface Phenomena Associated with Fouling and Cleaning in Food Processing*, (eds B. Hallstrom, D.B. Lund and Ch. Tragardh), Reprocentralen, Lund, Sweden, pp. 27–55.

Mackie, A.R., Mingins, J. and Dann, R. (1991) Preliminary studies of β-lactoglobulin adsorbed on polystyrene latex, in *Food Polymers, Gels and Colloids*, (ed. E. Dickinson), Special Publication No.82, Royal Society of Chemistry, Cambridge, pp. 96–112.

Marangoni, D.G., Smith, R.S. and Roscoe, S.G. (1989) Surface electrochemistry of the oxidation of glycine at Pt. *Can. J. Chem.* **67**:921–26.

Marangoni, D.G., Wylie, I.G.N. and Roscoe, S.G. (1991) Surface electrochemistry of the oxidation reactions of α- and β-alanine at a platinum electrode. *Bioelectrochem. Bioenerg.* **25**:269–84.

Roscoe, S.G. and Fuller, K.L. (1992) Interfacial behavior of globular proteins at a platinum electrode. *J. Coll. Interface Sci.* **152**:429–41.

Roscoe, S.G. and Fuller, K.L. (1993) An electrochemical study of the effect on temperature on the adsorption behavior of β-lactoglobulin. *J. Coll. Interface Sci.* **160**:243–51.

Rowley, B.O., Lund, D.B. and Richardson, T. (1979) Reductive methylation of β-lactoglobulin. *J. Dairy Sci.* **62**:533–36.

Sawyer, W.H. (1968) Denaturation of bovine β-lactoglobulins and relevance of disulfide aggregation. *J. Dairy Sci.* **51**:323–29.

Sivik, B. and Hallström, B. (1981) State-of-the-art of fouling of membrane surfaces, in *Fundamentals and Application of Surface Phenomena Associated with Fouling and Cleaning in Food Processing*, (eds. B. Hallström, D.B. Lund and Ch. Tragardh), Reprocentralen, Lund, Sweden, pp. 10–26.

Skudder, P.J., Brooker, B.E., Bonsey, A.D. and Alvarez-Guerrero (1986) Effect of pH on the formation of deposit from milk on heated surfaces during ultra-high-temperature processing, *J. Dairy Res.* **53**:75–87.

Tanford, C. (1968). Protein denaturation. *Adv. Protein Chem.* **23**:121–282.

Tissier, J.P., Lalande, M. and Corrieu, G. (1984) A study of milk deposit on heat exchange surfaces during UHT treatment, in *Engineering and Food*, Vol. 1, (ed. B.M. McKenna), Elsevier, London, pp. 49–58.

Walstra, P. and Jenness, R. (1984) *Dairy Chemistry and Physics*, Wiley-Interscience, New York, pp. 167–73.

Watanabe, K. and Klostermeyer, H. (1976) Heat-induced changes in sulphyhryl and disulfide levels of β-lactoglobulin A and the formation of polymers. *J. Dairy Res.* **43**:411–18.

Wong, D.W.S. (1989) *Mechanism and Theory in Food Chemistry*, Van Nostrand Reinhold, New York, p. 87.

8 Raman spectroscopy as a probe of protein structure in food systems

E. Li-Chan, S. Nakai and M. Hirotsuka

Abstract

Raman spectroscopy can be a useful tool to probe protein structure in solid and liquid food systems. Bands in the Raman spectrum arising from amide I, amide III and skeletal stretching modes of peptides and proteins are useful for characterizing backbone conformation, including the estimation of secondary structure fractions. Bands attributed to various stretching or bending vibrational modes of functional groups of amino-acid residues can be used to monitor the environment around these side-chains. In particular, valuable information may be obtained on SS or SH groups of cystinyl or cysteinyl residues, CH groups of aliphatic residues, and aromatic rings of tryptophanyl, tyrosinyl and phenylalanyl residues. One important parameter distinguishing Raman spectroscopy from many other spectroscopic methods is its applicability to systems containing high concentrations of proteins, which is critical for the investigation of structural changes during processes such as coagulum or gel formation. Thus, changes in both intramolecular and intermolecular interactions can be studied. In this chapter, examples are presented on the application of Raman spectroscopy to investigate protein structure as a function of processing, such as heating, drying, salt addition or homogenization with lipids, which may be important to correlate with protein functionality in food systems.

8.1 Introduction

Elucidation of the relationship between protein structure and function is the goal of protein chemists with diverse areas of interest. With the exponential growth in genetic and protein engineering techniques over the past decade, the ability to predict the function of a protein molecule that could result from intentionally engineered changes is even more crucial. While most studies on structure–function relationships have focused on peptides and proteins of physiological significance, that is, with 'biological activity', food scientists are also becoming aware of the importance of understanding the effect of protein structure on their functional properties in food systems, to enable tailoring of ingredients for specific uses.

Although numerous methods exist to study protein structure, few can be applied to investigate proteins under the conditions that are relevant to their eventual use in foods. Protein ingredients are usually not pure protein preparations. In addition to consisting of several protein constituents, various other components such as salts, lipids and carbohydrates are usually present in food systems. In addition, food proteins typically are converted from their native state to denatured or even insoluble forms by processing steps such as heating, drying, freezing or whipping.

Raman spectroscopy is a branch of vibrational spectroscopy that can give useful information on the vibrational motions of molecules. In essence, Raman spectroscopy is based on the shifts in wavelength or frequency of the exciting incident beam resulting from inelastic collisions with the sample molecules (Carey, 1982). Since both the intensity and frequency of molecular vibrations are sensitive to chemical changes and the environment around the atoms, the Raman spectrum can be used as a monitor of molecular chemistry. Distinct advantages of Raman spectroscopy over many other spectroscopic methods are its applicability to studying molecules in aqueous solutions, nonaqueous liquids, fibers, films, powders, gels and crystalline state. As a result of this versatility, it has been pointed out that Raman spectroscopy has a clear but as yet unrealized potential for characterizing the individual components of food systems and should be sensitive to structural changes induced by processing such as mixing, aeration, heating, fiber-formation and gelation (Painter, 1984).

The aim of this chapter is to introduce Raman spectroscopy to protein chemists who may not have previously considered it as a tool to study food proteins. A brief history and introduction of the Raman effect and instrumentation are presented first, followed by a review of typical band assignments and interpretation of protein spectra. Some examples are shown to illustrate the applications of this method to probe changes in protein molecules as a function of processing such as heat denaturation, gelation and emulsification. Finally, limitations of the present technique, recent trends in instrumentation and data-handling, and potential future applications to studying protein structure–function relationships in food systems are discussed.

8.2 Overview of Raman spectroscopy

8.2.1 Basis for Raman spectroscopy

Infra-red and Raman spectroscopy are both techniques that can be classified under the broader term, 'vibrational spectroscopy'. The former involves the absorption of radiation in the infra-red region of the electromagnetic spectrum, whereas the latter involves the inelastic scattering of radiation, usually in the visible region but also in the ultraviolet and near-infra-red regions.

The relationships between infra-red absorption, elastic (Rayleigh) scattering and inelastic (Raman) scattering are depicted in Figure 8.1a. Absorption and emission bands involve discrete vibrational transitions in the ground electronic state. Absorption of energy corresponding to $h\nu_i$ is measured in infra-red absorption spectroscopy. Vibrational energy transitions are also involved during scattering of light. In the case of Rayleigh scattering, the frequency of the scattered radiation is identical to the incident radiation ν_0, whereas in the case of Raman scattering the frequency of scattered radiation is shifted $\pm\, \nu_i$ from the incident radiation, resulting in Stokes $(\nu_0 - \nu_i)$ lines and anti-Stokes $(\nu_0 + \nu_i)$ lines. In Raman spectroscopy, vibrational transitions are measured in terms of the shift in frequency or wavenumber (cm^{-1}), ν_i, from the incident wavenumber, ν_0, as shown in Figure 8.2. Rayleigh scattered light is not shifted from the incident radiation and thus appears at a wavenumber shift of 0 cm^{-1}.

In Figure 8.1b is depicted the basis of the phenomena responsible for a branch of Raman spectroscopy known as resonance Raman spectroscopy. The distinction between resonance and nonresonance Raman scattering lies in the energy of the incident or exciting beam. Ultraviolet lasers are usually used for resonance Raman spectroscopy, while visible lasers are used for normal Raman spectroscopy. For resonance Raman scattering to occur, the incident photon energy must be of the order of the electron transition energy, sufficient to allow excitation to an excited electronic state, followed by inelastic scattering to produce Stokes and anti-Stokes resonance Raman lines. It is important to note that the bands or lines resulting in both Raman and resonance Raman spectra are a property of the electronic ground state. Figure 8.1b also shows how fluorescence may result when an excited sample molecule loses part of its energy through nonradiative transition and drops to the lowest level in the excited state, prior to emitting fluorescence during the process of returning to the ground state.

Both infra-red absorption and Raman scattering processes involve the vibrational energy levels of sample molecules, which are related primarily to stretching or bending deformations of bonds. However, infra-red absorption requires a change in the intrinsic dipole moment with molecular vibration, while Raman scattering depends on changes in the polarizability of the molecule. Thus, the two methods are complementary rather than substitutes for each other. Polar functional groups such as C=O, C≡N and C–H usually have strong infra-red stretching vibrations, whereas intense Raman lines are associated with nonpolar groups such as C=C, S–S and N=N (Painter, 1984). Water has a strong infra-red absorption band and interferes with interpretation of aqueous sample spectra but it is a poor Raman scatterer, therefore it is a suitable solvent for Raman spectroscopy.

8.2.2. Raman instrumentation and sample-handling techniques

The Raman effect was first discovered by Sir C.V. Raman (1928) and Raman and Krishnan (1928), who noted that liquid irradiated with blue light, produced by cutting off other lights with a filter, emitted greenish light. However, Raman scattering is inherently very weak compared with elastic or Rayleigh

(a)

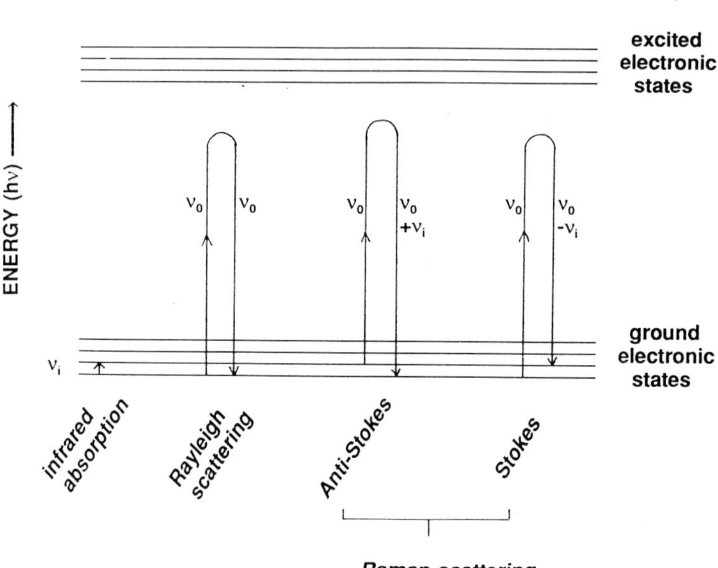

(b)

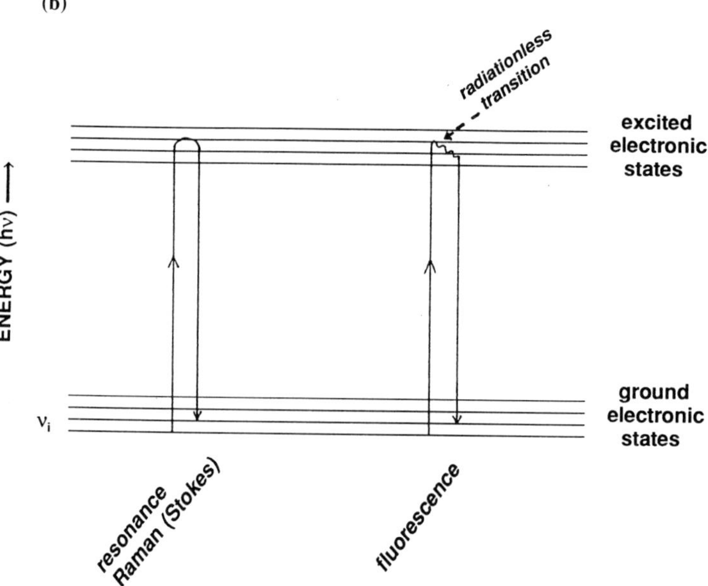

Figure 8.1 Schematic presentation of: (a) infrared absorption, Rayleigh scattering, and Stokes and anti-Stokes lines from Raman scattering; (b) resonance Raman scattering and fluorescence phenomena.

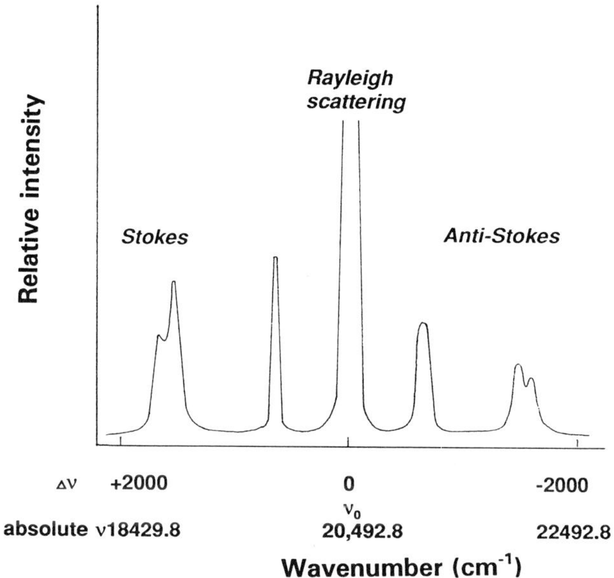

Figure 8.2 Typical Raman spectrum showing the shift in wavenumber from the incident radiation (e.g. argon ion laser excitation at 488 nm, absolute wavenumber = 20 492.8 cm^{-1}).

scattering. Thus, Raman spectroscopy only became a routine laboratory tool after the advent of lasers capable of producing intense monochromatic radiation suitable for excitation.

The basic layout of a typical Raman spectrophotometer is shown in Figure 8.3. Intense monochromatic radiation, usually provided by continuous wave gas lasers such as argon or krypton ion gas, is focused on the sample. The resulting scattered light is gathered by collection optics and directed to a dispersing system, typically a double or triple monochromator, which spatially separates the scattered light on the basis of frequency. The Raman spectrum is then detected and recorded either sequentially by a single photomultiplier used with a scanning monochromator, or simultaneously by a multichannel detector. In most cases, modern instruments are interfaced with computers or microprocessors to collect and store the data for further analysis.

Many variations of the basic design exist. For example, the range of wavelengths available for Raman excitation can be extended and made nearly continuous through the use of a frequency doubling system pumped by ion lasers or dye lasers. In particular, the virtual elimination of fluorescence problems has been reported by using near-infra-red Fourier transform (NIR-FT) Raman spectrometers, which employ Nd:YAG (neodymium:yttrium aluminium garnet) laser radiation at 1064 nm for excitation of Raman spectra and interferometers, with Fourier transformation of the interferograms (Góral and Zichy, 1990; Schrader *et al.*, 1991). Photon

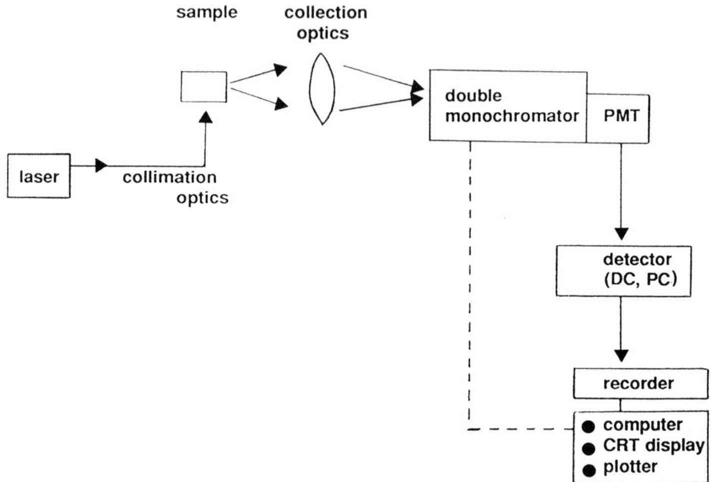

Figure 8.3 Schematic presentation of a basic Raman spectrometer: PMT, photomultiplier tube; DC, direct current amplification; PC, photon counting; CRT, cathode ray tube.

counting systems can be used instead of direct current amplification systems for recording the Raman signal from the photomultiplier tube. Further details of instrumentation can be found in the literature (e.g. Yu, 1977; Carey, 1982; Strommen and Nakamoto, 1984).

Many variations also exist with respect to sample-handling techniques. Since solids (including crystals, powder, films and fibres), liquids and gases can all be investigated using Raman spectroscopy, numerous different sample-holder sizes and geometries are either available commercially or can be constructed according to the needs of individual investigators. Perhaps the most commonly used, especially for nonresonance Raman spectroscopy, is a plain glass or hematocrit capillary tubing since only small quantities of sample are required; the ends can be sealed if desired for hygroscopic, volatile, corrosive or oxygen-sensitive samples, and the glass does not absorb Raman scattered light in the visible region. Although glass exhibits Raman scattering, slit height can be adjusted to introduce only scattered light from the sample to the entry slit of the monochromator. Ninety degree scattering is typically employed for transparent samples, whereas 180° scattering (or 'backscattering') is often recommended for opaque samples. In some cases, rotating sample holders, flow-through cells or thermostatically controlled cells are used to prevent overheating and decomposition of sample by the laser, and for conducting studies at varying temperatures (Carey, 1982; Strommen and Nakamoto, 1984).

8.3 Raman spectroscopy of proteins

Proteins and their components have the distinction of being the classical example of application of Raman spectroscopy to biomolecules (Carey, 1982). Early work on amino acids, peptides and proteins was carried out using mercury lamps as the excitation source, which resulted in very weak lines in the Raman spectrum due to protein scattering. Lord and Yu (1970) published the first laser Raman spectrum of a protein, lysozyme, in aqueous solution. Assignments of the bands were carried out by comparing the protein spectrum with that obtained by a mixture of the constituent amino acids. Since that time, a multitude of different biomolecules, including proteins, have been studied by Raman spectroscopy, and a growing body of knowledge has been accumulated to aid in interpretation of the Raman spectra of proteins (Frushour and Koenig, 1975; Yu, 1977; Carey, 1982; Painter, 1984; Alix *et al.*, 1985; Hudson and Mayne, 1986; Tu, 1986; Harada and Takeuchi, 1986; Krimm, 1987; Yager and Gaber, 1987; Bertoluzza *et al.*, 1989). In addition to knowledge gained from experimental data, techniques of normal-mode analysis have been applied for prediction of theoretical locations of bands in the Raman spectra of oligopeptides and polypeptides (Krimm and Bandekar, 1986; Bandekar and Krimm, 1988).

In this chapter, emphasis will be placed on information that can be obtained from protein spectra by nonresonance or classical laser Raman spectroscopy. Although resonance Raman spectroscopy has the advantages of enhancement of Raman scattering, ability to use relatively dilute sample solutions and potential to reduce the problem of fluorescence from chromophores that absorb in the visible region (Hudson and Mayne, 1986), in practice, laboratories are more likely to be equipped with lower-cost lasers such as argon, krypton or helium-neon lasers, which have usable lines in the visible spectrum (Yager and Gaber, 1987). The concentrations of protein naturally found in many food systems are usually high enough so that the Raman scattering effect observed by visible laser excitation is strong enough to yield intense bands in the spectrum. In fact, the low protein concentrations used for resonance Raman spectroscopy would not be comparable with the levels actually encountered in most food systems. However, the problem of fluorescence may sometimes be a serious one in foods, especially those that contain pigments such as carotenoids in plant foods or heme in products from animal origin. Although various solutions to the fluorescence problem have been suggested, including sample purification, prolonged exposure to the laser beam, collision quenching by iodide addition or by variations of instrumental technique (e.g. Yu, 1977; Carey, 1982; Strommen and Nakamoto, 1984), in many cases alternatives to conventional laser Raman spectroscopy may be the only solution. This will be illustrated later in this chapter by an example of the use of NIR-FT-Raman compared with conventional visible laser Raman spectroscopy to study soy protein denaturation, gelation and protein–lipid interactions.

8.3.1 Side-chain vibrations

Table 8.1 shows the assignments of bands or lines that may be seen in the Raman spectrum of proteins, to vibrational motions of the side-chains of various amino-acid residues. In general, the locations of these bands are similar for proteins in solid form (e.g. lyophilized or crystalline) or dissolved in water, although small shifts that may be related to changes in the protein structure in the different states are sometimes observed. Dissolution in heavy water (D_2O) also leads to shifts in some bands compared with the spectrum in H_2O, as discussed below. Some of the bands that can give information about the microenvironment around the side-chains are described in the following sections.

8.3.1.1 Sulfur-containing amino acids.

Raman spectroscopy is a useful tool to detect the presence of disulfide bonds due to the intense band assigned to the disulfide (S–S) stretching vibration in the region of 500–550 cm^{-1}. Most naturally occurring proteins and peptides that contain disulfide bonds show a band near 510 cm^{-1} in the Raman spectrum. This corresponds to the lowest potential energy conformation of the disulfide bond, that is, 'gauche-gauche-

Table 8.1 Typical wavenumber of Raman bands and general assignments for side-chain vibrations in Raman protein spectra[a]

Wavenumber (cm^{-1})	Assignment or interpretation
	aliphatic amino acid residues
2800–3000	C–H stretch
1465, 1450	CH_2 bend
	polar or charged residues
1700–1750	C=O stretch of Asp, Glu COOH
1400–1430	C=O stretch of Asp, Glu COO⁻
1650	–CO–NH– of Asn
1615	–CO–NH– of Gln
1640, 1600	NH_3^+ of Lys
1491	imidazole of His
1399	OH of Ser, Thr (weak)
	aromatic amino acid residues
3050–3100	C–H stretch
1605, 1585, 1207, 1030, 1006, 622	Phe
1610, 1590, 1263, 1210, 1180, 850, 830, 645	Tyr
1622, 1582, 1553, 1363, 1014, 879, 761, 577, 544	Trp
	sulfur-containing residues
540	trans-gauche-trans
525	gauche-gauche-trans } SS of cystine
510	gauche-gauche-gauche
745–700	trans C–S of Met, Cys, Cys/2
670–630	gauche C–S of Met, Cys, Cys/2
2550–2580	S–H of cysteine

[a]Source: adapted from Li-Chan and Nakai, 1991a.

gauche' conformation of C–C–S–S–C–C. Additional bands appearing at 525 and 540 cm^{-1} have been assigned to the '*gauche-gauche-trans*' and '*trans-gauche-trans*' conformations, respectively. This interpretation of the dependence of the S–S stretching vibration on the internal rotation about the C–S and C–C bonds was first proposed by Sugeta *et al.* (1972, 1973) and supported by various workers (e.g. Nakanishi *et al.*, 1974; Kitagawa *et al.*, 1979; Byler *et al.*, 1983). A different theory was proposed by Van Wart *et al.* (1973) and Maxfield and Scheraga (1977) in which the S–S stretching band is related to the C–C–S–S dihedral angle; the bands at 525 and 540 cm^{-1} were postulated to arise from presence of conformations with small C–C–S–S dihedral angles. As Tu (1986) summarized, interpretation of the disulfide bond vibrational wavenumber in published papers is divided according to these two theories, although recent evidence supports the interpretation of Sugeta *et al.* (1972, 1973).

Stretching vibration due to the S–H group of cysteinyl residues appears in the vicinity of 2550–2580 cm^{-1}, similar to that for many organic compounds. The detection of the sulfhydryl group is usually simplified by the absence of any other vibrational bands in this region of the protein spectrum. Further confirmation can be made by analyzing the Raman spectrum of the deuterated protein, since the conversion of S–H to S–D causes an isotopic shift in the stretching vibration to about 1865 cm^{-1}, the theoretical ratio of νS–H/νS–D being $\sqrt{2}$ or 1.4.

C–S stretching vibrations from Met, Cys and 1/2 Cys residues of proteins generally appear in the 600–750 cm^{-1} region. The vibrational wavenumber appears to depend on the conformation in the environment of the C–S bond. For example, the *trans* form of methionine shows C–S stretching bands at 655 and 724 cm^{-1}, whereas in the *gauche* form the band appears at 700 cm^{-1}.

8.3.1.2 Tyrosine. Several Raman bands have been assigned to Tyr-residue vibrational modes, including those around 650, 830, 850, 1180, 1200–1210, 1265–1290, 1500–1510 and 1600–1620 cm^{-1}. This amino-acid residue often plays an important role in hydrogen bond formation or as a co-ordination ligand for metals in metalloproteins (Tu, 1986).

Among the many Raman bands, the pair located at 830 and 850 cm^{-1} are most useful for monitoring the microenvironment around Tyr residues. These doublet bands arise from Fermi resonance of the ring-breathing vibration and an overtone of an out-of-plane ring-bending vibration of the para-substituted benzene ring (Siamwiza *et al.*, 1975). Two interpretations have been proposed for these bands. The first is that the intensity ratio depends on whether the Tyr residue is exposed or buried; the second is that the ratio depends on the state of hydrogen bonding of the phenolic OH group. In fact, the two interpretations may be related, since when the Tyr residue is exposed, its hydroxyl group can interact with solvent and act as a simultaneous acceptor and donor of moderate to weak hydrogen bonds, while a buried Tyr residue tends to act as a hydrogen bond donor to the carboxylate ion of Asp or Glu residues (Figure 8.4). In the former

(a)   **(b)**

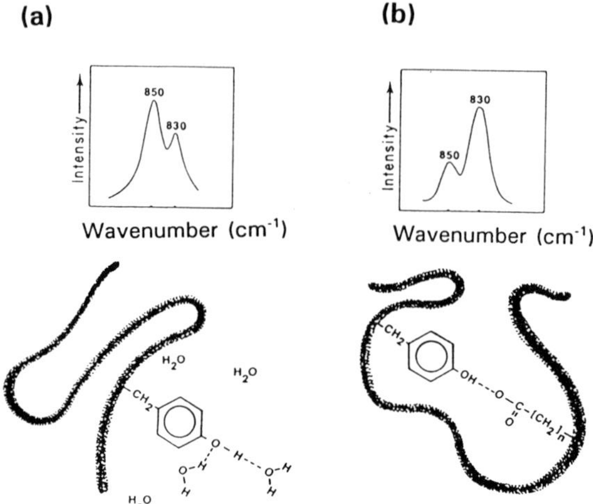

Figure 8.4 Relative intensity of the 850 and 830 cm^{-1} doublet in the Raman spectrum for: (a) exposed tyrosine residues, which may act as hydrogen bond acceptor or donor; and (b) buried tyrosine residues, which act as hydrogen bond donor. (Adapted from Tu, 1986.)

case, the intensity ratio $I_{850/830}$ tends to be high (0.9–1.45), whereas in the latter case the ratio is lower, usually between 0.7 and 1.0. However, the ratio can be as low as 0.3 in the case of extremely strong hydrogen bonding to a negative acceptor (Van Dael *et al.*, 1987) or as high as 2.5 for the case of a phenolic OH group, which acts as a strong hydrogen-bond acceptor (Carey, 1982). The doublet has been used to determine the number of buried and exposed Tyr residues in proteins using the following equations (Tu, 1986), where N refers to mole fraction:

$$N_{buried} + N_{exposed} = 1 \qquad (8.1)$$

$$0.5N_{buried} + 1.25N_{exposed} = I_{850/830} \qquad (8.2)$$

8.3.1.3 Tryptophan. Many of the Raman bands attributed to Trp residues are derived from indole-ring vibrations. Tryptophan vibrations for lysozyme were noted at 544, 577, 761, 879, 1014, 1338, 1363, 1553 and 1582 cm^{-1} (Lord and Yu, 1970). Generally, indole rings in buried or hydrophobic environments give rise to sharp identifiable peaks in protein Raman spectra near these wavenumbers (Carey *et al.*, 1986). Various workers have proposed using the intensity of the 1363, 880 or 760 cm^{-1} bands, or the trough between the 1340–1360 cm^{-1} doublet, to monitor the buried or exposed state of Trp residues (Kitagawa *et al.*, 1979; Itoh *et al.*, 1983; Tu, 1986; Miura *et al.*, 1991). A band at 1386 cm^{-1} from deuterated Trp residues has been reported to be useful to follow the kinetics of H–D exchange and the rate of exposure of Trp residues to the solvent (Carey, 1982).

8.3.1.4 Phenylalanine. This aromatic amino-acid residue also shows several bands in protein Raman spectra, including a strong band at 1003–1006 cm^{-1} attributed to the breathing vibration of the monosubstituted ring. Weak bands are observed at 620 and 1030 cm^{-1} and various bands appear at 1180, 1200, 1585 and 1610 cm^{-1}, but these bands often overlap with those from Tyr residues. The intensity of the Phe bands does not appear to be influenced by the micro-environment or external factors (Tu, 1986), and the strong band at 1003–1006 cm^{-1} may thus serve as a useful internal standard (e.g. Barrett *et al.*, 1978; Lippert *et al.*, 1981; Harada *et al.*, 1982).

8.3.1.5 Histidine. Histidine residues can be detected by a band in the 1490–1500 cm^{-1} region. However, this band does not seem to be sensitive to changes in the environment or state of the imidazole ring, and often is masked by vibrational bands of other amino-acid residues such as Tyr. Upon denaturation, a strong Raman band has been observed at 1410 cm^{-1}, which has been assigned to N$_1$–C$_2$–N$_3$ symmetric stretching and N–D bending vibrations of the imidazole ring. The band appears to be sensitive to change from imidazolium to imidazole and to be useful to determine the ionization states of His residues in proteins (Lord and Yu, 1970; Harada *et al.*, 1982).

8.3.1.6 Aspartic and glutamic acids. The carboxyl group vibrational bands can be used to monitor state of ionization since the ionized group (COO$^-$) exhibits a band at 1400–1420 cm^{-1} while the undissociated form (COOH) exhibits bands at 1700–1750 cm^{-1}. By comparing the intensities of the protein Raman spectrum in these two regions, one can estimate relative ionization state of the carboxyl groups (Tu, 1986).

8.3.1.7 Aliphatic amino acids. C–H stretching and deformation bands of aliphatic amino acids appear in the 2800–3000 and 1400–1500 cm^{-1} regions, respectively. The C–H stretching vibrational bands include fundamental C–H stretching vibrations, overtones of various C–H stretching vibration and Fermi resonance between C–H stretching and C–H deformation bands. Due to the overtone and resonance interactions, many bands appear in the 2800–3000 cm^{-1} region. Basically, bands at 2900, 2940 and 2980 cm^{-1} have been assigned to CH$_2$ symmetric, CH$_2$ asymmetric and CH$_3$ asymmetric stretching vibrations, respectively. Verma and Wallach (1977a) reported that unfolding of ribonuclease A was accompanied by a large increase in the intensity of the 2930 cm^{-1} band, while Mikkelsen *et al.* (1978) observed increased intensity of this band resulting from a transmembrane potential and/or cation gradient across erythrocyte vesicle membranes. These workers interpreted amplification of the 2930 cm^{-1} band to reflect transfer of protein CH$_3$ groups from previously buried or apolar environment to the surrounding aqueous or polar environment.

Although the C–H stretching vibrations occur in a region of the Raman spectrum in which there are few other bands assigned to protein functional groups, this region also contains C–H vibrational bands from lipids (Verma and Wallach, 1977b). However, the location of the bands are usually sufficiently different enough to distinguish their origins (Tu, 1986). This region thus has been suggested to be useful for studying protein–phospholipid interaction, membrane proteins, and protein–lipid interactions in emulsions (Larsson and Rand, 1973; Larsson, 1976; Mikkelsen *et al.*, 1978; Lippert *et al.*, 1981; Aslanian *et al.*, 1986; Tu, 1986; Carmona *et al.*, 1987; Yager and Gaber, 1987; Li-Chan and Nakai, 1991a).

Most C–H deformation bands are found in the region of 1400–1500 cm^{-1}, and are sensitive to carbon-chain configuration (Tu, 1986). However, structural implications of these bands have not been extensively studied or reported. Many investigators have used the 1445–1450 cm^{-1} methylene H–C–H or C–H deformation band as an internal standard for protein spectra, assuming that its intensity is relatively independent of protein structural changes (Lord and Yu, 1970; Barrett *et al.*, 1978; Caillé *et al.*, 1983; Tu, 1986). Lippert *et al.* (1976) and Arêas *et al.* (1989) suggested that possible information could be obtained on the interaction between the hydrophobic C–H moiety and its environment by measuring a 'C parameter', defined as the relative intensity of methylene deformation observed in proteins at 1448 cm^{-1} compared with the intensity in poly-L-lysine. According to Lippert *et al.* (1976) the C parameter tended to increase with a decreasing amount of solvent. Therefore, it was inferred that for proteins in an aqueous or hydrophilic medium, high values of the C parameter would result if the hydrophobic CH_2 groups were exposed, while lower values would result in the case of increased hydrophobic interactions between CH_2 groups.

8.3.2 *The peptide backbone*

The –CO–NH– amide or peptide bond has several distinct vibrational modes, including amide I, II, III, IV, V, VI, VII, A and B bands, as shown in Table 8.2. In practice, many of these bands are weak or absent in the Raman spectra of proteins, and only amide I and amide III bands, which have relatively high intensity in

Table 8.2 Peptide bond vibration assignments, based on the model compound N-methylacetamide[a]

Wavenumber (cm^{-1})	Assignment
3280	Amide A
3090	Amide B
1653	Amide I
1567	Amide II
1299	Amide III
627	Amide IV
725	Amide V
600	Amide VI
206	Amide VII

[a]Source: adapted from Yu, 1977.

the Raman spectra are used to investigate protein structure or conformation. Hydrogen bonding usually decreases the amide I band wavenumber but increases the amide III band wavenumber.

Owing to the possible overlap of solvent water Raman bands in the amide I region, and of miscellaneous C–H bending and aromatic ring vibrations in the amide III region, it is recommended that both amide I and amide III regions be inspected before making any conclusions about structure or conformational changes. Additional information obtained upon deuteration or from bands other than the amide I and III bands should also be considered.

8.3.2.1 Amide I band. The amide I band arises primarily from in-plane peptide C=O stretching vibrations, and partly from in-plane N–H bending vibrations. The exact location of the amide I band in the Raman spectrum depends on hydrogen bonding and conformation of the polypeptide or protein molecule. Model studies have been used to assign the amide I bands for synthetic polypeptides such as α- and β-forms of poly-Gly, poly-L-Ala and poly-L-Lys, and these have been used to aid in the interpretation of the protein amide I bands.

The ranges for band wavenumbers typically found for the different protein conformations (α-helix, β-sheet, β-turn and random coil) are shown in Figure 8.5. Generally, proteins with high α-helical content show an amide I band centred around 1645–1657 cm^{-1}, while those with predominantly β-sheet structures show the band at 1665–1680 cm^{-1}. Proteins containing a high proportion of undefined or random coil structure have an amide I band close to 1660 cm^{-1}. In fact, because many proteins contain various secondary structure fractions, the Raman spectrum often shows several components or shoulders in this region.

8.3.2.2 Amide III band. The amide III band primarily results from C–N stretching and N–H in-plane bending vibrations of the peptide bond, and is located in the 1200–1300 cm^{-1} region. As in the case of the amide I band, assignments of location of the band as a function of protein conformation have been aided by use of synthetic polypeptides of known structure, and protein spectra often show several component bands in this region.

The typical ranges of the amide III bands for the different protein conformations are shown in Figure 8.5. β-sheet structures are usually readily detected around 1238–1245 cm^{-1}, whereas random coil structures appear near 1250 cm^{-1}. The intensity of the amide III band for α-helix structures is usually weak or moderate, and is located in a broad region from 1260–1300 cm^{-1}, which overlaps with the region assigned for β-turns. Owing to the weaker intensity of the helical conformation, helix to random coil or helix to β-sheet transitions are often accompanied by an increase in intensity and broadening around the amide III region.

Deuteration of the peptide bond to form –CO–ND– results in a shift in the amide III band. The ratio for the wavenumber shift caused by deuteration, v_{NH}/v_{ND}, has been reported to be 1.27–1.29 for various proteins (Tu, 1986),

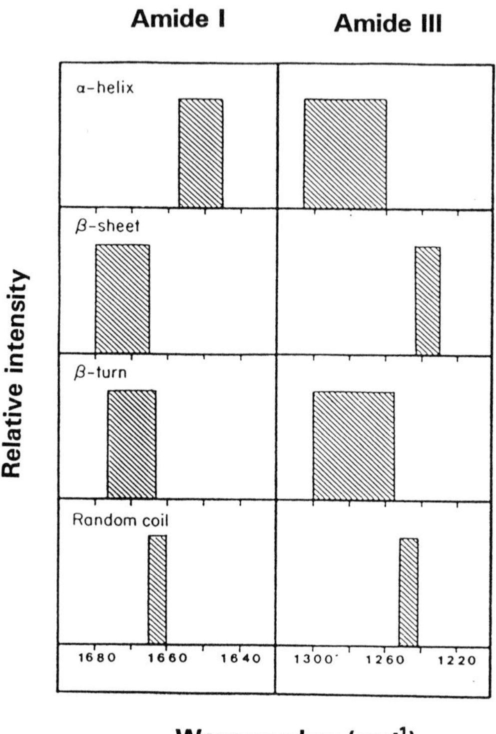

Figure 8.5 Typical location of Amide I and III vibrations corresponding to different types of protein conformation. (Adapted from Tu, 1986.)

compared with the theoretical value of 1.4. The deuterated amide III' band is usually located at 950, 980 and 1012 cm^{-1} for α-helix, antiparallel β-sheet and parallel β-sheet structures, respectively.

8.3.2.3 Other bands. In addition to the characteristic location of the amide I and III bands, α-helical conformation is often accompanied by the appearance of a band in the 890–945 cm^{-1} region, arising from skeletal C–C stretching vibrations, while β-sheet structures often show a band in the 1020–1060 cm^{-1} region. The detection of these other bands can be used to support the findings based on the amide I and III bands.

8.3.2.4 Estimation of secondary structure. Over the past 15 years, vibrational spectroscopic characteristics, especially of the amide I and amide III bands, have been used not only for qualitative investigation of the different classes of protein conformation but also for quantitative analysis, thereby permitting estimation of the relative proportions of the conformational classes in individual proteins. Two broad categories of methods exist for estimation of secondary structure, namely theoretical prediction methods and physicochemical techniques (Yada *et al.*, 1988). Prediction methods are based on the amino-acid sequence of the protein, and attempt to give information on locations of various conformational states such as α-helix, β-sheet and β-turn. Of the physicochemical techniques, X-ray diffraction or crystallographic analysis is currently the most accurate method for determining three-dimensional structure, including secondary structure locations, of a protein in crystalline form. Circular dichroism, in conjunc-tion with curve-fitting techniques, is the most powerful analytical method for estimating secondary structure fractions of proteins in solution. However, the former technique is often limited by constraints of time and cost, as well as difficulty in obtaining suitable crystals. Circular dichroism is only suitable for analysis of nonturbid or nonparticulate solutions. Serious discrepancies between estimates of α-helix, β-sheet and random coil content by circular- dichroism and by X-ray diffraction have been reported for various proteins; it was suggested that this might be due to conformational changes accompanying dissolution (Lippert *et al.*, 1976). Lower content of α helical structure and/or higher content of β sheet structure have been reported when comparing crystalline versus solution states of proteins. In this respect, vibrational spectroscopic techniques including infra-red and Raman spectroscopy hold promise to clarify changes in conformation arising from differences in the state of the protein, including crystallization, lyophilization, precipitation and dissolution in various solvents. However, it must be emphasized that these methods cannot indicate the locations of structural features but can only give an estimation of the amounts of various secondary structures and how these may change under varying conditions.

Various methods have been proposed for quantification of secondary structural content of proteins from their Raman spectra. Lippert *et al.* (1976) proposed solving a set of simultaneous equations involving spectral heights at 1240 cm^{-1} in H$_2$O (amide III) and at 1632 and 1660 cm^{-1} in ^{2}H$_2$O (amide I'), measured relative to the 1448 cm^{-1} methylene band. Williams (1981, 1983, 1986) developed methods whereby the amide I and III bands were analyzed as linear combinations of the corresponding bands in spectra of proteins in a reference set whose secondary structures were known *a priori*. A 'Reference Intensity Profiles' method was proposed by Berjot *et al.* (1987) based on the principle of computation of Raman spectral intensities corresponding to pure classes of secondary conformations giving rise to peptide backbone amide I vibrations. The computations were performed by a step-by-step fit of experimental spectra of a large number of proteins whose secondary structures in crystal form were well known from X-ray diffraction data. The group of Byler, Susi and co-workers adapted their experience in Fourier transform infra-red spectroscopy and second derivative spectroscopy for conformational analysis (Susi and Byler, 1983, 1987, 1988a; Byler and Susi, 1986) and applied Fourier deconvolution of the amide I Raman band to estimate secondary structure of casein and various other food proteins (Byler and Susi, 1988; Byler *et al.*, 1988; Susi and Byler, 1988b).

8.4 Examples of applications of Raman spectroscopy to probe protein structure in food systems

Raman spectroscopy has been used mainly as a probe of protein structure in biochemistry (Yu, 1977), with expanding applications in the areas of membrane protein and lipid biochemistry (e.g., Yager and Gaber, 1987; Vogel and Jähnig, 1986; Van de Ven *et al.*, 1984; Carmona *et al.*, 1987), medical research (Ozaki, 1988) including the study of cataract formation using intact lens (Itoh *et al.*, 1983) as well as lyophilized and solution states of lens proteins (Pande *et al.*, 1989; Chen *et al.*, 1991), comparison of solid versus solution states of proteins (e.g. study of hen egg yolk phosvitin by Prescott *et al.*, 1986), crystalline forms of proteins (e.g. avidin by Honzatko and Williams, 1982; proteins in the endotoxin crystals from *Bacillus thuringiensis* by Carey *et al.*, 1986) and even intact egg shell (e.g. eggshell proteins from the fish *Salmo gairdneri* by Hamodrakas *et al.*, 1987). Proceedings from two recent European conferences on spectroscopy of biological molecules (Alix *et al.*, 1985; Bertoluzza *et al.*, 1989) show diverse applications of Raman spectroscopy. In this chapter, a few examples have been chosen to illustrate possible avenues of research in applying this technique to probe protein structure that is important for function in food systems.

8.4.1 *Effects of salts or ions on protein structure*

8.4.1.1 Muscle systems. The effects of the salts $CaCl_2$, $MgCl_2$, and LiBr on Raman spectra of aqueous solutions of myosin were studied by Barrett *et al.* (1978). Analysis of the Raman bands in the regions at 900, 940, 1240–1300 and 1650–1670 cm^{-1} led to the interpretation that the Ca^{2+} ion effects an α-helix to β-sheet transition in myosin, while LiBr appeared to denature the protein, resulting in increased random coil structure. Magnesium chloride was reported to have an effect intermediate between the other two salts. These authors speculated that an increase in β-structure would mean an increase in interchain hydrogen bonding, which is compatible with the observation that 0.1 mM Ca^{2+} induces reversible aggregation of myosin heads.

Raman spectra of intact muscle fibers and internally perfused fibers in capillary tubes were reported by Caillé *et al.* (1983). The use of internally perfused fibers ensured good control of the concentration of Ca^{2+}, Mg^{2+} and ATP. By using a spectral subtraction technique, it was observed that the contraction that occurred upon either removal of ATP or Mg^{2+} or addition of Ca^{2+} was accompanied by a marked decrease in the intensity of the carboxylate band at 1414 cm^{-1} and of the Trp band at 758 cm^{-1}, while bands associated with the amide I band at 1650 cm^{-1} and C–C stretching at 940 cm^{-1} increased in intensity. These results were interpreted as being due to strong electrostatic interactions between basic and acidic residues during contraction and to changes in α-helical content or orientation of some of the contractile proteins.

8.4.1.2 Salt-induced precipitation or 'salting out'. The molal surface tension increment of a salt, σ, was proposed by Melander and Horvath (1977) to be useful in classifying salts as salting in or salting out types. Salts with greater denaturing potential and salting out efficiency have smaller σ values. The effects of various types and amounts of salts (Na_2SO_4, NaCl, NaBr, KBr, and KSCN) on secondary structure of the resulting salt-induced precipitates of α-chymotrypsin were studied by Raman spectral analysis of the amide I region (Przybycien and Bailey, 1989). As the molal surface tension increment of the inducing salt decreased, β-sheet content of the salt-induced precipitate increased while α-helix content decreased. The secondary structure contents did not appear to depend, however, on the amount of the salts used.

The contents of α-helix and β-sheet of twelve different proteins in their native state and in precipitates obtained using the chaotropic salt KSCN versus the structure stabilizing salt Na_2SO_4 were also estimated from analysis of the amide I band in Raman spectra (Przybycien and Bailey, 1991). Statistical analysis of the estimated perturbations in the secondary structure contents indicated the formation of β-sheet structures with a concomitant loss of α-helix on precipitation with KSCN. These results were further investigated with respect to primary structure in terms of fractions of hydrophobic and charged amino acids, and it was suggested that hydrophobic forces dominate changes in the β-sheet while electrostatic forces are operative in changes in the α-helix. Based on the correlation of these observations on salting-out phenomena with reports from other investigators on formation of β-sheet-like structure in proteins and peptides under various environments and aggregative processes, these authors hypothesized that the formation of β-sheet strands may be a fundamental phenomenon in self-associating or aggregating protein systems.

8.4.2 Denaturation, aggregation and gelation

8.4.2.1 Egg-white proteins. Raman spectral shifts in the conformationally sensitive amide I and amide III regions were noted upon heating whole egg white as well as ovalbumin (Painter and Koenig, 1976). The appearance of an intense amide III line at 1236 cm^{-1} and a shift in the amide I from 1667 to 1672 cm^{-1} were interpreted as formation of stable intermolecular β-sheet structures, which were deemed of central importance in the thermal denaturation and aggregation of egg white. Involvement of new disulfide bridges was discounted due to absence of any new bands appearing in the S–S stretching region near 500 cm^{-1}.

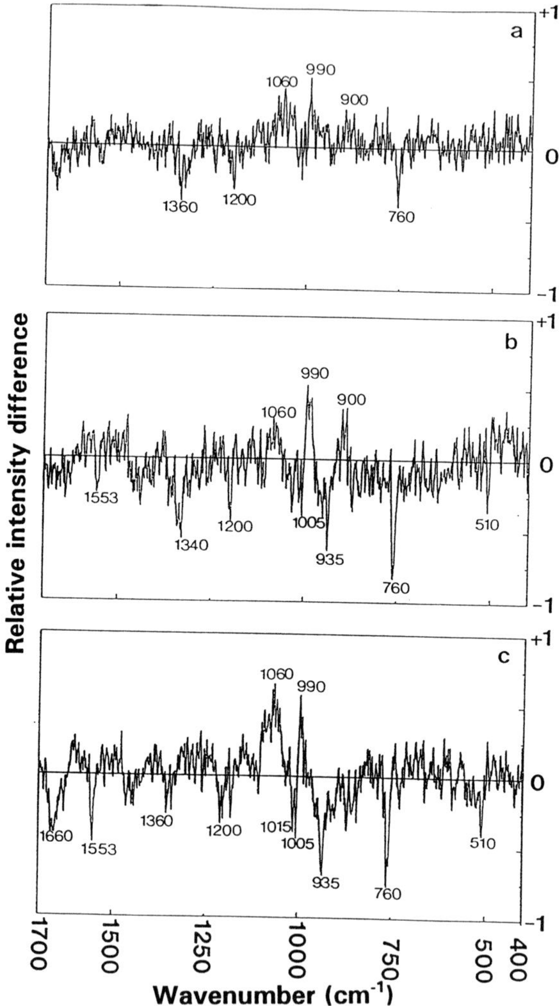

Figure 8.6 Raman difference spectra of 20% lysozyme solutions in D_2O, as follows: (a) [lysozyme, 10 mM DTT, 40°C, 10 min] − [lysozyme, 10 mM DTT]; (B) [lysozyme, 10 mM DTT, 75°C, 5 min] − [lysozyme, 10 mM DTT]; and (c) [lysozyme, 100°C, 5 min] − [lysozyme]. (From Li-Chan and Nakai, 1991b.)

Raman and infra-red spectroscopy were used to investigate the structure of aggregates and gels of lysozyme (10 mg/ml) formed in acidified alcohol solutions; comparisons were made with the structure of the soluble forms of the protein at concentrations of 0.01–4 mg/ml studied by circular dichroism (Nemoto *et al.*, 1983). It was reported that an intense peak at 1238 cm^{-1} in the gel may be related to antiparallel β-sheet, and two new lines at 1303 and 1316 cm^{-1} in the gel could be assigned to additional β-turns. Decreases in the intensity in the 1260–1280 and 933 cm^{-1} regions were considered to reflect a decrease in α-helical content in the gels. Analysis of the interaction parameters between intra- and intermolecular peptide groups from the amide I' mode suggested that intermolecular interactions played an important role in stabilization of the gel. In contrast, lysozyme in 6 M GuHCl showed a broad amide III band with several small peaks, indicative of random coil conformation. Circular dichroism results indicated that at a low protein concentration of 0.01 mg/ml the protein remained as a monomer with α-helix rich conformation, while at 1 and 4 mg/ml β-sheet structures became dominant.

The effects of heating hen egg lysozyme in the absence or presence of low or high concentrations of a reducing agent, dithiothreitol (DTT), were investigated by Raman spectroscopy (Li-Chan and Nakai, 1991b). In the absence of reducing agent the Raman spectrum of lysozyme was relatively unchanged by heating at temperatures up to 75–80°C. However, in the presence of 10 or 100 mM DTT, opaque gels were formed even by heating under mild conditions such as at 37 or 60°C. Some of the changes resulting from these treatments are illustrated in Figure 8.6, which shows the difference spectra of gelled minus solution states of the lysozyme obtained under different conditions. A decrease in α-helical content and an increase in anti-parallel β-sheet content for molecules in gels compared with those in solution were suggested by the negative and positive peaks in the difference spectra at 935–950 and 980–995 cm^{-1}, respectively. This trend was also confirmed by quantitative estimation of the secondary structure fractions by analysis of the amide I region using a modified method of Williams (1983) as reported by Przybycien and Bailey (1989).

Denaturation and gelation were also accompanied by increased exposure of aromatic residues indicated by decreased intensity of the bands at 760, 880 and 1340–1360 cm^{-1}. Changes in the disulfide bond conformation were noted in S–S stretching bands at 509–530 cm^{-1}, but appearance of a band at 2580 cm^{-1}, indicating generation of sulfhydryl groups, was observed only under some conditions. Intermolecular disulfide cross-links were not always necessary for gelation. Reduction or interchange reactions involving the intramolecular disulfide bonds, which normally stabilize the native tertiary structure of lysozyme, were proposed as being required to allow molecular flexibility, thus allowing conformational changes and exposure of hydrophobic groups leading to gel formation. The reactions leading to this destabilization could be brought about either by high heat treatment, or by low concentrations of DTT in conjunction with mild heating conditions.

8.4.2.2 Muscle systems. Raman spectra of myosin extracted from freshly excised rabbit psoas muscle showed an intense feature at 1244 cm^{-1}, suggestive of either β-structure or random coil, as well as a 1265 cm^{-1} shoulder associated with the globular myosin head and a 1304 cm^{-1} shoulder attributed to the fibrous helical conformation of the myosin tail region (Carew *et al.*, 1975). Increasing the temperature from 20 to 68°C increased the intensity of the 1244 cm^{-1} band, which might be expected upon denaturation to random coil or upon increasing β-sheet structures; such an increase might be involved in myosin head aggregation. Thermal denaturation was also reported to lead to changes in the 1040–1120 cm^{-1} region, which contains conformationally sensitive skeletal vibrations—especially C–N and C–C stretching modes.

8.4.2.3 Bovine serum albumin. Denaturation of bovine serum albumin by heating, acid or alkali treatment was investigated by Raman spectroscopy by Lin and Koenig (1976). Heating to 70°C, or change in the pH of the solution to below pH 5 or above pH 10, caused gradual intensification of the 1246 cm^{-1} band relative to 1337 cm^{-1}, and a decrease in intensity of the 938 cm^{-1} band relative to 1003 cm^{-1} (Figure 8.7). Drying the gel formed at 70°C in air for a week led to formation of a clear solid, which showed a shift in the amide III band from 1246 to 1240 cm^{-1}. The Raman spectrum of bovine serum albumin gel formed

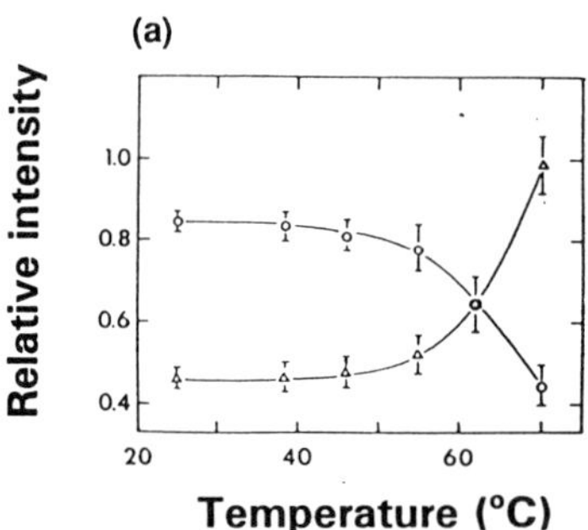

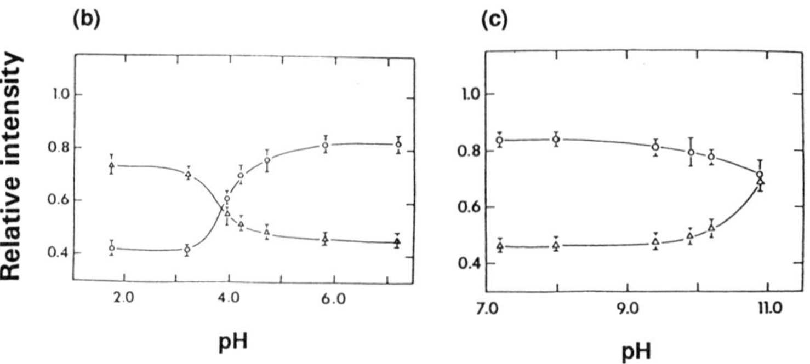

Figure 8.7 Changes in the intensity at 938 and 1246 cm^{-1} relative to 1003 cm^{-1} upon denaturation of bovine serum albumin by: (a) heat, (b) acid or (c) alkali treatment: O, I_{938}/I_{1003}; △, I_{1246}/I_{1337}). (Adapted from Lin and Koenig, 1976.)

by heating at 90°C for 30 min also showed an intense line at 1240 cm⁻¹. These results were interpreted as unfolding of α-helices upon denaturation, which were observed as the decreased intensity at 938 cm⁻¹, and accompanied by an increase in disordered structure reflected in the 1246 cm⁻¹ band. These changes were reversible at temperatures up to about 50°C as well as for acid or alkali induced denaturation between pH 1.72 and 10.9. At higher temperatures that led to aggregation and gel formation, β-structures were formed as reflected by appearance of a band at 1240 cm⁻¹.

Similar observations of decreased α-helical content and increased β-sheet content were made in thermally induced gels of bovine serum albumin in deuterium oxide, by measurements of the infra-red amide I' region as well as the amide I, amide III and 940 cm⁻¹ regions in Raman spectra (Clark *et al.*, 1981). These authors reported that no obvious differences in secondary structure were observed for gels of varying clarity and texture but quantitative estimation of the secondary structure fractions was not carried out to support this statement. It was suggested that generation of sheet structure during aggregation may be a general phenomenon, although some proteins such as ribonuclease do not follow this trend.

8.4.3 Comparison of solution versus solid phases

8.4.3.1 Bovine serum albumin. The Raman spectra of bovine serum albumin in crystalline state and as a 5% solution in water were reported to differ in several regions (Lin and Koenig, 1976). The solid-state protein showed the amide I band at 1658 cm⁻¹, consistent for a protein with significant α-helical conformation. Dissolution caused a shift to 1652 cm⁻¹ and an intensification of the 935 cm⁻¹ band, which was interpreted as an adjustment of distorted to more regular helix structures. Changes in the acidic residues comprising 22% of the amino-acid composition were also suggested by the appearance of a band at 1733 cm⁻¹ in the solid state, compared with a band at 1402 cm⁻¹ in solution; these bands were assigned to carbonyl stretching of the undissociated and ionized forms, respectively, of carboxyl groups.

8.4.3.2 Ribonuclease. Changes in the intensity of ring vibrations of tyrosine residues of ribonuclease A were suggested to be due to conformational changes resulting from the dissolution of ribonuclease A into aqueous solution (Yu *et al.*, 1972). Upon dissolution at pH 8.89, a decrease in intensities of the Raman bands at 644 and 852 cm⁻¹ relative to the intensity at 832 cm⁻¹ was observed, which was suggested to be related to change in local environment of the three buried Tyr residues of ribonuclease; these residues have an abnormally high pK_a value of greater than 11. Changes upon dissolution were also noted in the amide III region. In the powder form, a line at 1239 cm⁻¹ was assigned to β-structure while a line at 1260 cm⁻¹ was assigned to α-helices. At pH 8.89, the 1260 cm⁻¹ line shifted to 1265 cm^{-1} and both lines sharpened. Sharpening and amplification of a band at 937 cm^{-1} may also have been related to increased α-helical structure.

In contrast, dissolution of the ribonuclease powder at pH 1.5 resulted in an increase in the relative band intensity at 644 and 852 cm^{-1} compared with that at 832 cm^{-1}, which could be interpreted as exposure of the buried Tyr residues to an aqueous environment. Again, changes in the conformation were indicated by an amplification of the 1265 cm^{-1} band and a marked decrease in the band at 937 cm^{-1}.

8.4.3.3 Hen egg yolk phosvitin. Phosvitin in neutral aqueous solution shows an amide I band at the unusually high wavenumber of 1682 cm^{-1} with a shoulder at 1668 cm^{-1}, while the amide III region contains a strong band at 1253 cm^{-1} with a weak shoulder at 1284 cm^{-1} (Prescott *et al.*, 1986). These features, together with the spectral characteristics in deuterium oxide and the rapid and complete deuterium exchange of amide NH, were interpreted as indications of deficiencies in both ordered α-helical and β-sheet structures of conventional geometry. In other words, at physiological pH this protein was suggested to contain an open, solvent-accessible structure.

Large changes in secondary structure, particularly significant contents of β-sheet structure indicated by an amide I band at 1672 cm^{-1} and an amide III band at 1237–1242 cm^{-1}, were observed when phosvitin was either dissolved at acidic pH (below 2) or was lyophilized from a neutral aqueous solution. With lyophilization, the amide I and III bands assumed the pattern observed for the β-sheet structure of low pH phosvitin, even though the effective pH was still neutral after lyophilization, as judged by bands related to phosphoserine side-chain ionizations.

8.4.4 Protein–lipid interactions

Raman spectroscopy can be used to study interactions between protein and lipid components, especially by monitoring changes in the C–H stretching vibrations, which can be observed in the 2800–3000 cm^{-1} region (Table 8.3). However, most of the published literature in this area has been on phospholipid–protein

Table 8.3 C–H Stretching vibration of protein and lipid components[a]

Assignment of vibration	Wavenumber (cm^{-1})	
	Protein	Lipid
CH$_2$ symmetric	2900	2850
CH$_2$ asymmetric	2940	2885
CH$_3$ asymmetric	2980	2930

[a]Source: Li-Chan and Nakai, 1991a.

interactions related to biological membrane structures (e.g. Aslanian *et al.*, 1986; Carmona *et al.*, 1987; Yager and Gaber, 1987). Rather limited information has been published on the study of protein–lipid interactions in emulsion systems. Some of these investigations are reviewed below.

Changes in the environment of hydrocarbon chains of different lipids and phosphate esters as well as application to lipid–protein systems were studied by Larsson and Rand (1973) and Larsson (1976). However, these studies were focused on the changes in the lipid components, and it was assumed that the contribution of protein components to the spectrum was insignificant. They reported that although the region between 1000–1200 cm^{-1} attributed to C–C stretching vibration could give information on *trans* and *gauche* conformation of the hydrocarbon chains, the 2800–3000 cm^{-1} region related to C–H stretching is more sensitive to structural changes in lipid–water and lipid–protein–water phase systems. The ratio of the intensity of peaks at 2850 and 2885 cm^{-1}, corresponding to symmetric and asymmetric stretching vibrations of the CH_2 groups, respectively, is related to the disorder of the hydrocarbon chains; the I_{2850}/I_{2885} ratio increases with the degree of 'looseness' of the lateral packing of disordered hydrocarbon chains in the following order: lamellar liquid crystal < hexagonal or cubic liquid crystal < micellar solution < solution in an organic solvent. Thus, the peak at 2850 cm^{-1} is expected to be dominant in the liquid state of the hydrocarbon chains, whereas the peak at 2885–2890 cm^{-1} should dominate when the hydrocarbon chains are in crystalline form. The latter case was observed for the fat globules in bovine milk, suggesting that most of the hydro-carbon chains of lipids at the oil–water interface or milkfat globule membrane in the milk 'emulsion' are crystalline and closely packed. In contrast, the peak at 2850 cm^{-1} was found to be dominant in the corresponding milk fat separated in the form of a plastic fat (Larsson, 1976).

In the presence of various proteins or polypeptides, the relative intensity of a peak at 2930 cm^{-1}, attributed to asymmetric stretching vibrations of CH_3 groups, appeared to increase with increased degree of polarity of the environ-ment around the hydrocarbon chains. It was suggested that this could be used to detect whether the hydrocarbon chains of the lipid molecules are associated into separate lipid regions or located in a protein environment (Larsson and Rand, 1973).

The change in the intensity of the 2930 cm^{-1} band of ribonuclease relative to that of the 2850 cm^{-1} CH_2 stretching band of egg lecithin was used to monitor changes in the environment of the aliphatic amino-acid residues (Verma and Wallach, 1977b). It was reported that marked amplification near 2930 cm^{-1} may reflect exposure of apolar protein side-chains to an aqueous milieu. Use of the 2850 cm^{-1} band of lecithin as an internal standard was claimed to be justified owing to its thermal stability over the range of temperature used to monitor ribonuclease unfolding; however, there may be changes in this band due to ribonuclease–lecithin interaction upon heating.

Recently the authors reported Raman spectra of emulsions formed with lysozyme, or its partially reduced derivative, and corn oil (Li-Chan *et al.*, 1991a).

Figure 8.8 shows the Raman spectra in the 2500–3150 cm^{-1} region for lysozyme and DTT-treated lysozyme solutions, and corn oil and emulsions formed by vortex action or sonication. Changes in the relative intensity and wavenumber of bands in the C–H stretching region were observed, which could be due to interactions of protein and lipid components. For example, decrease in intensity of scattering at 2930–2940 cm^{-1} suggested a decrease in the exposure of protein

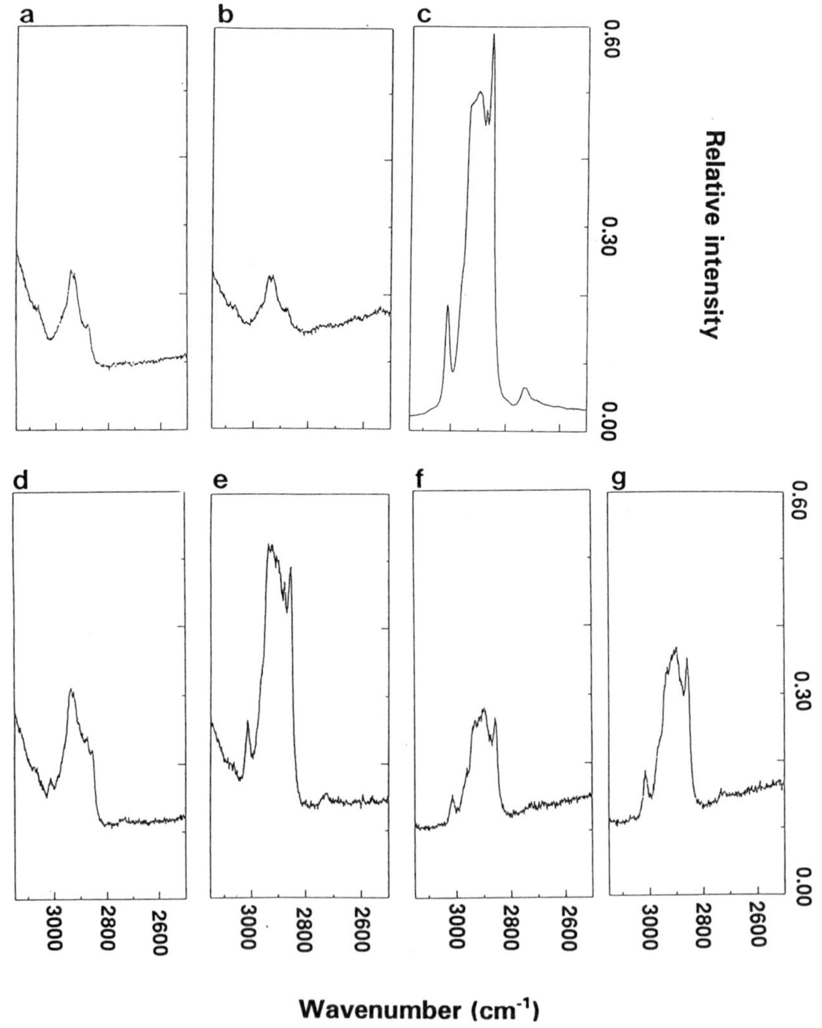

Figure 8.8 Raman spectra in the 2500–3150 cm^{-1} region of: (a) 10% lysozyme solution, pH 7.25; (b) 10% lysozyme + 10 mM DTT, pH 7.25; (c) corn oil; and their emulsions (ø=0.25) formed by (d) and (e) vortex action (30 s) or (f) and (g) sonication (15 watts, 10 min at 50% pulse rate in an ice bath).

aliphatic side chains to an aqueous environment, possibly due to interaction with lipid molecules. A shift in the wavenumbers of some bands in the Raman spectrum of the emulsion was observed by subtracting the individual spectra of protein and corn oil from the emulsion (Figure 8.9). The decrease in intensity at 2849, 2875, 2930 and 3005 cm^{-1} and the increase in intensity at 2858, 2895, 2970 and 3015–3020 cm^{-1} in these difference spectra suggested interactions leading to greater ordering or immobilization of the fatty acid chains, similar to changes reported in the spectra of lipids undergoing a liquid to solid transition (Verma and Wallach, 1977a).

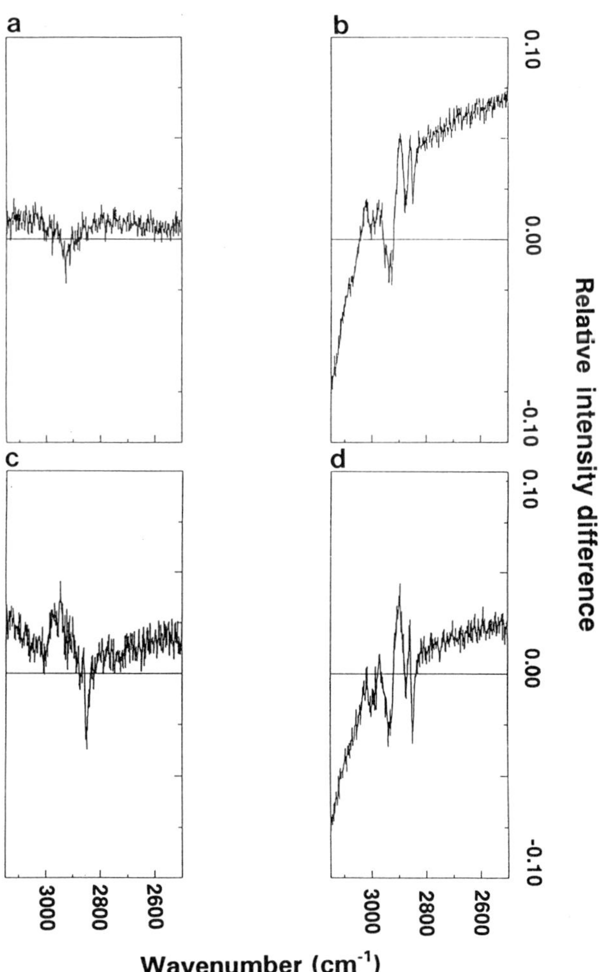

Figure 8.9 Raman difference spectra obtained by subtraction of corn oil and protein spectra from emulsion spectra of: (a) vortexed lysozyme; (b) sonicated lysozyme; (c) vortexed DTT-lysozyme; and (d) sonicated DTT-lysozyme emulsions prepared as described in the legend to Figure 8.8. (Adapted from Li-Chan & Nakai, 1991a.)

Using this approach, further work is being carried out to investigate protein–lipid interactions in emulsions formed using different food proteins, for example, β-lactoglobulin, β-casein, bovine serum albumin, and egg albumen proteins (E. Li-Chan, 1991, unpublished data). Two examples are illustrated below.

Figure 8.10a shows the difference spectrum obtained by subtracting the individual spectra of β-casein and corn oil from the spectrum of the emulsion formed by vortex action. A small negative peak at 2850 cm^{-1} suggested some immobilization of the hydrocarbon chains from the oil component. A more extensive difference was seen in the difference spectrum for the cream layer obtained after leaving the emulsion at ambient temperature for 5 h (Figure 8.10b).

The Raman spectra of egg white, corn oil and the emulsion formed by vortex action are shown in Figure 8.11, while Figure 8.12 shows the difference spectrum for the egg white emulsion, obtained by subtracting the spectrum of: (i) the corn oil component, (ii) egg white component, and (iii) both corn oil and egg white. The difference spectra thus obtained represent Raman bands resulting from protein plus interactions (Figure 8.12a), corn oil plus interactions (Figure 8.12b) and only the interactions (Figure 8.12c) resulting from

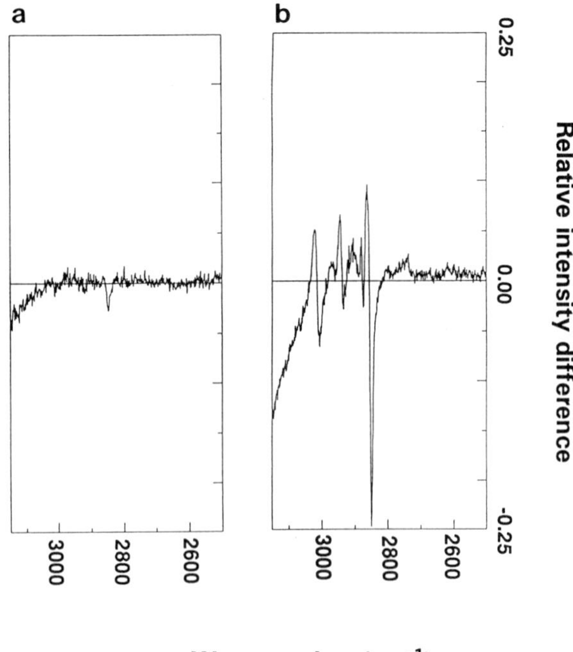

Figure 8.10 Raman difference spectra obtained after subtraction of corn oil and β-casein spectra from the spectra of: (a) the emulsion (ø=0.25), formed by vortex action; and (b) the cream layer obtained after standing of the emulsion for 5 h at ambient temperature.

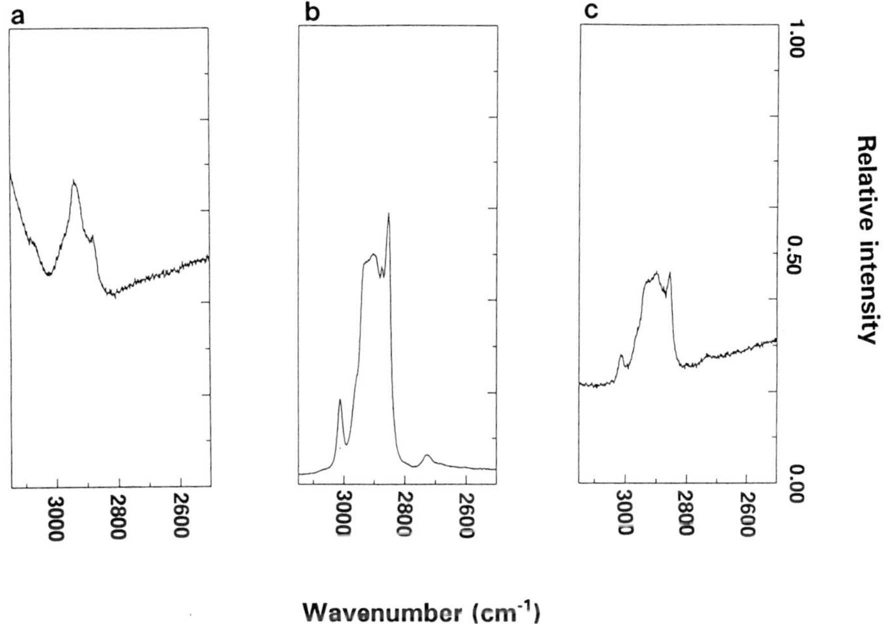

Figure 8.11 Raman spectra of: (a) egg white; (b) corn oil and (c) their emulsion (ø=0.25) formed by vortex action.

emulsification. The negative peak at 2850 cm^{-1} (Figure 8.12a), as well as the sharpening and shifts in bands, attributed to the corn oil (Figure 8.12b compared with Figure 8.11b) indicated a change in the hydrocarbon chains to a more 'solid-like' or immobilized state in the emulsion. Shifts in the protein C–H stretching bands were also noted, for example, the band at 2942 cm^{-1} and shoulder at 2879 cm^{-1} in the egg white spectrum (Figure 8.11a) shifted to the two bands at 2945 and 2900 cm^{-1}, respectively, in the emulsion (Figure 8.12a).

8.5 Recent trends in Raman spectroscopy

Many advances in Raman spectroscopic techniques have evolved in the past decade. Since Raman spectra of complicated systems, such as protein solutions, often consist of broad and complex band contours, many attempts have been made to improve interpretation of these spectral data by applying various curve-fitting or resolution enhancement techniques. In particular, with the advent of Fourier transform spectroscopy, Fourier convolution and deconvolution transform methods have been used widely for improving quality of the Raman spectral data. Ni and Scheraga (1985) suggested that a maximum entropy method (MEM) was especially useful for resolution enhancement of the complicated Raman spectra of proteins, compared with the conventional discrete Fourier transform (DFT) method. deNoyer and Dodd (1990, 1991) recently recommended a statistical probability based method known as maximum likelihood estimation as being preferable to Fourier deconvolution methods. Derivative Raman spectroscopy has also been suggested for facilitating interpretation (Van de Ven *et al.*, 1984).

In addition to resolution enhancement techniques, various intensity enhancement methods have also been proposed. Many examples of resonance Raman spectroscopy have appeared in the literature in which strong Raman signals are observed by laser excitation in the ultraviolet region, including the probing of buried Tyr and Trp residues (Caswell and Spiro, 1986), study of intermediates in enzyme substrate complexes (Bajdor *et al.*, 1987) and many other applications (Harada and Takeuchi, 1986; Hudson and Mayne, 1986). Surface-enhanced Raman spectroscopy (SER) and surface-enhanced resonance Raman spectroscopy (SERRS), in which molecules are adsorbed onto certain metal surfaces such as silver, copper and gold, lead to enhancement in Raman scattering which can be on the order of 10^9 (Chumanov *et al.*, 1990; Nabiev *et al.*, 1990).

Fluorescence continues to be a problem in Raman spectroscopic analysis of

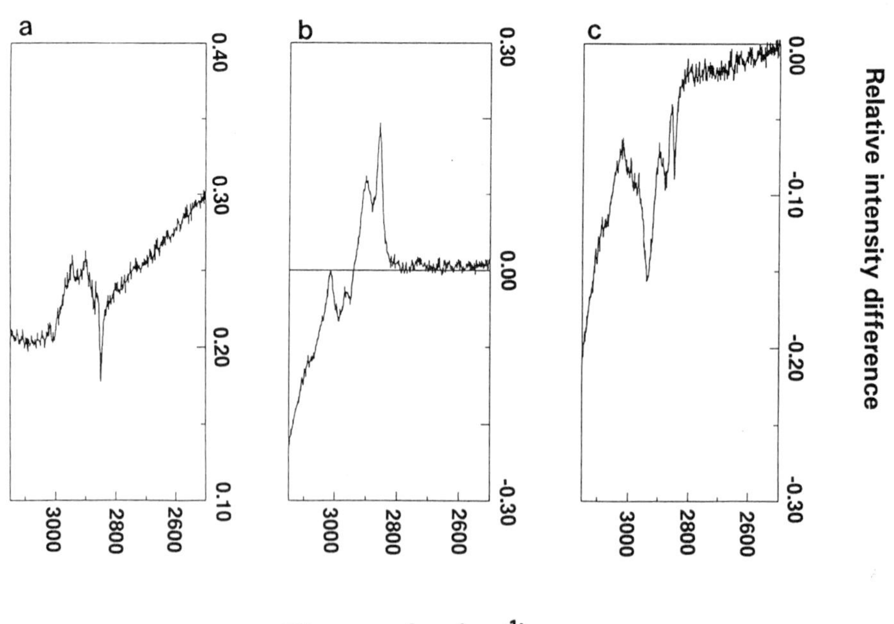

Figure 8.12 Raman difference spectra obtained from the spectra shown in Figure 8.11 as follows: (a) emulsion − corn oil; (b) emulsion − egg white; and (c) emulsion − corn oil − egg white.

many biological samples when visible laser excitation is used. In some cases, use of ultraviolet region excitation can significantly decrease this interference. Nonlinear techniques such as coherent anti-Stokes Raman spectroscopy (CARS) have also been used successfully. As indicated in the review by Gerrard and Bowley (1988), a variety of different methods have met with some degree of success, although sample fluorescence often remains a problem.

A significant advance in solving the fluorescence problem for many samples has been the development of NIR-FT-Raman spectroscopy. The use of diode-laser pumped Nd:YAG laser radiation at 1064 nm in conjunction with inter-ferometers and Fourier transform have resulted in several distinct advantages, namely: (i) virtual elimination of fluorescence; (ii) good signal-to-noise ratio; and (iii) markedly shortened data-collection times. These are illustrated in

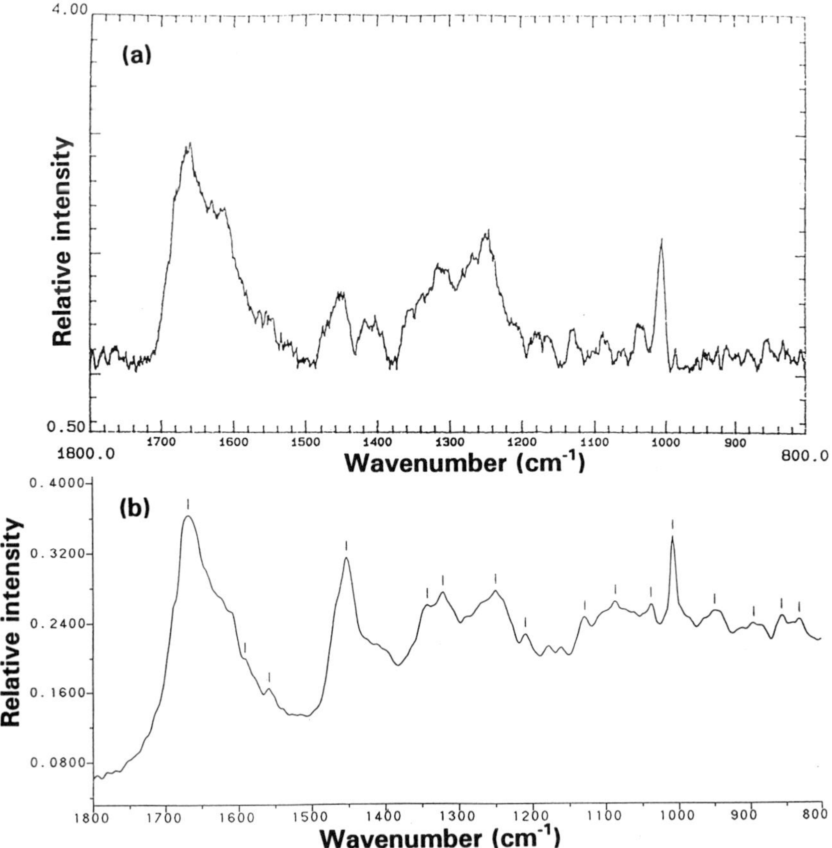

Figure 8.13 Raman spectra of soy 7S globulin obtained by: (a) visible laser Raman spectroscopy using argon ion laser excitation at 488 nm with multichannel detector: and (b) NIR-FT Raman spectroscopy using Nd:YAG laser excitation at 1064 nm with InAsGe detector and interferometer.

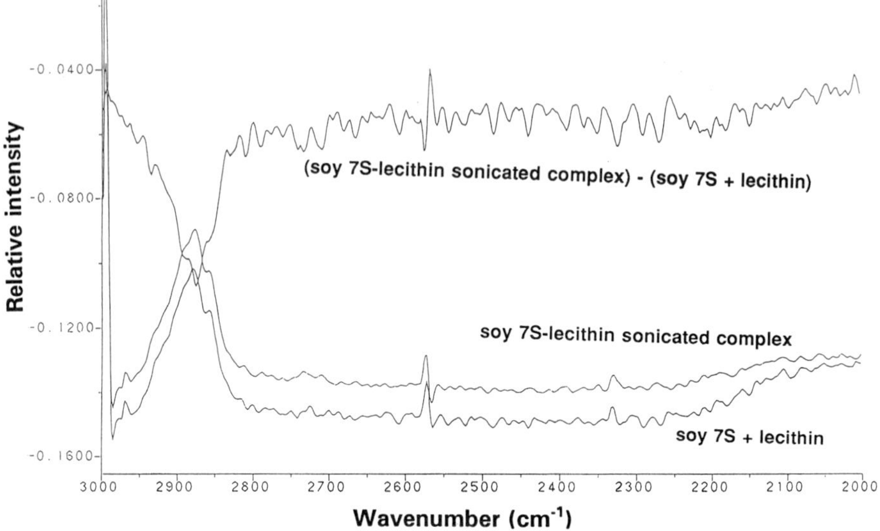

Figure 8.14 NIR-FT Raman spectra and difference spectrum of soy 7S globulin-soy lecithin sonicated complex compared with a non-interacting mixture of the two components.

several recent reviews (Góral and Zichy, 1990; Levin and Lewis, 1990; Schrader
et al., 1991), as well as in an example of its application to the study of lens
protein (Chen *et al.*, 1991).

The advantage of NIR-FT-Raman over conventional visible laser Raman spectroscopy has also been well demonstrated in our experience on the study of
soybean proteins, including effects of heating, gelation and pressure, as well as in
the study of protein–phospholipid interactions. Figure 8.13 illustrates the typical
signal–noise ratio of the spectrum of a 10% aqueous solution of soy 7S globulin,
obtained from conventional laser Raman spectroscopy using argon ion laser
excitation (Figure 8.13a) compared with that from NIR-FT-Raman using
Nd:YAG laser excitation (Figure 8.13b). Using visible laser Raman spectroscopy,
changes were observed in various peaks, attributed to protein–phospholipid interactions between 7S soy globulin with soy lecithin in a sonicated mixture
(Hirotsuka and Nakai, 1990; Nakai *et al.*, 1991). However, an intense fluorescence from this sample yielded low signal to noise and fluctuations in scattering
intensity during collection of the spectra, resulting in an inability to make any
detailed interpretations.

By using excitation at 1064 nm coupled with a large number of scans obtained
using NIR-FT-Raman technique, fluorescence as well as signal-to-noise problems were both resolved. Figures 8.14 and 8.15 show results obtained using this
technique to study soy protein–lecithin interactions, protein structural changes
during heating, gelation and high-pressure treatment. Figure 8.14 shows the
difference spectrum obtained by subtracting that of a noninteracting mixture of
soy 7S globulin and soy lecithin from the complex formed by sonication. A large

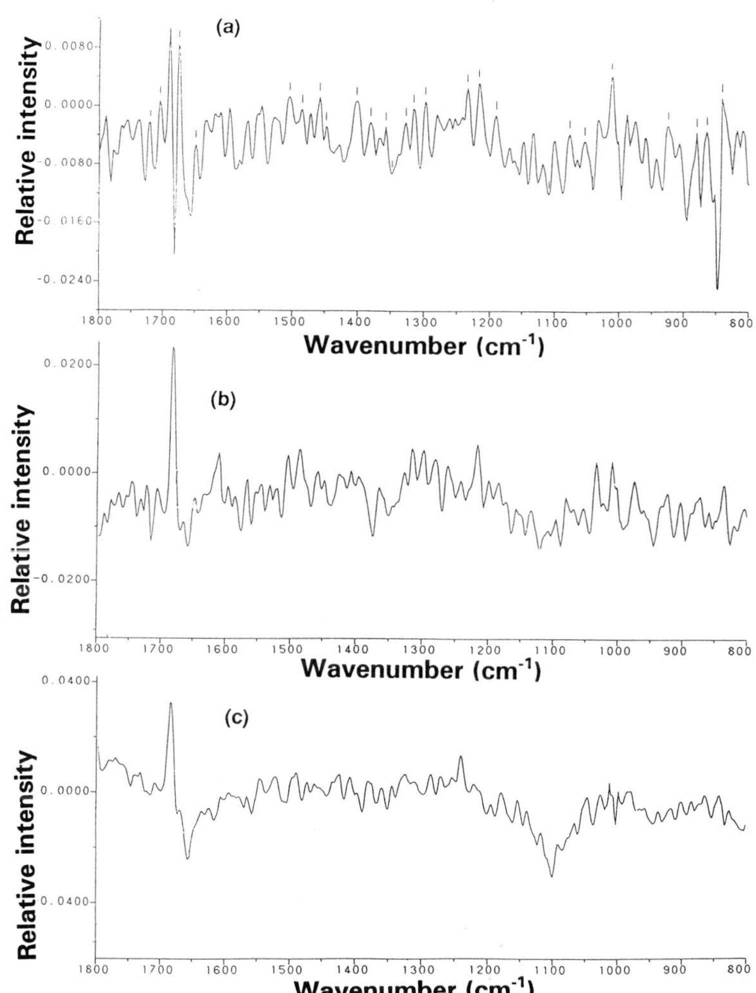

Figure 8.15 NIR-FT Raman difference spectra showing effects on soy 7S globulin of heating, high-pressure treatment and sol-to-gel transition, as follows: (a) [soy 7S, 100°C, 5 min] – [soy 7S]; (b) [high pressure-treated soy 7S] – [soy 7S]; and (c) [soy 7S gel] – [soy 7S solution].

negative band can be seen in the C–H stretching region, centred at 2875 cm^{-1}, with shoulders at 2850 and 2885 cm^{-1}; a positive band near 2570 cm^{-1} may suggest change(s) in the sulfhydryl groups of the protein in the sonicated mixture. Figure 8.15 illustrates that conformational changes can be detected in the amide I region near 1650–1680 cm^{-1} when soy protein is subjected to heating at 100°C, high-pressure treatment, as well as sol to gel transition. Changes in the tyrosine doublet near 830 and 850 cm^{-1} are also clearly observed, especially in the case of samples heated at 100°C.

The NIR-FT technique has limitations, however, including the inability to observe Raman bands near the existing wavelength because of the requirement of a secondary filter in that region. Nevertheless, its application to a variety of 'raw materials' such as nucleic acids, proteins, polysaccharides and fats in food products has been demonstrated by Góral and Zichy (1990). As Schrader *et al.* (1991) pointed out:

> the NIR-FT technique closes a gap; together with the well established Raman techniques with dispersive single- and multi-channel detectors and the non-linear Raman effects, all techniques serve to apply Raman spectroscopy to every problem where it could be appropriate; spectroscopists now have all the tools required to bring about the Raman renaissance.
>
> (Schrader *et al.*, 1991)

The challenge lies ahead to apply these techniques to the study of food systems, including the relationship between structure and function of food proteins.

References

Aslanian, D., Négrerie, M. and Chambert, R. (1986) A Raman spectroscopic study on the interaction of an ion-channel protein with a phospholipid in a model membrane system (gramicidin A/L-α-lysophosphatidylcholine). *Eur. J. Biochem.* **160**:395–400.

Alix, A.J.P., Bernard, L. and Manfait, M. (1985) *Spectroscopy of Biological Molecules*, Wiley Press, New York.

Arêas, E.P.G., Laure, C.J., Gabilan, N., Araujo, P.S. and Kawano, Y. (1989) Raman and infrared studies on the conformation of porcine, pancreatic and *Crotalus durissus terrificus* phospholipases A_2. *Biochim. Biophys. Acta* **997**:15–26.

Bajdor, K.., Peticolas, W.L., Wharton, C.W. and Hester, R.E. (1987) Ultraviolet (239 nm) resonance Raman spectroscopy of an enzyme-substrate intermediate of papain. *J. Raman Spectrosc.* **18**:211–14.

Bandekar, J. and Krimm, S. (1988) Normal mode spectrum of the parallel-chain β-sheet. *Biopolymers* **27**:909–21.

Barrett, T.W., Peticolas, W.L. and Robson, R.M. (1978) Laser Raman light-scattering observations of conformational changes in myosin induced by inorganic salts. *Biophys. J.* **23**:349–58.

Berjot, M., Marx, J. and Alix, A.J.P. (1987) Determination of the secondary structure of proteins from the Raman amide I band: the reference intensity profiles method. *J. Raman Spectrosc.* **18**:289–300.

Bertoluzza, M., Fagnano, C. and Monti, P. (1989) *Spectroscopy of Biological Molecules. State of the Art*, Proceedings of the 3rd European Conference. Esculapio, Bologna.

Byler, D.M. and Susi, H. (1986) Examination of the secondary structure of proteins by deconvolved FTIR spectra. *Biopolymers* **25**:469–87.

Byler, D.M. and Susi, H. (1988) Application of computerized infrared and Raman spectroscopy to conformation studies of casein and other food proteins. *J. Ind. Microb.* **3**:73–88.

Byler, D.M., Susi, H. and Farrell, H.M. Jr (1983). Laser-Raman spectra, sulfhydryl groups, and conformation of the cystine linkages of β-lactoglobulin. *Biopolymers* **22**:2507–11.

Byler, D.M., Farrell, H.M. Jr and Susi, H. (1988) Raman spectroscopic study of casein structure. *J. Dairy Sci.* **71**:2622–29.

Caillé, J.-P., Pigeon-Gosselin, M. and Pézolet, M. (1983) Laser Raman study of internally perfused muscle fibers. Effect of Mg^{2+}, ATP and Ca^{2+}. *Biochim. Biophys. Acta* **758**:121–27.

Carew, E.B., Asher, I.M. and Stanley, H.E. (1975) Laser Raman spectroscopy—new probe of myosin substructure. *Science* **188**:933–35.

Carey, P.R. (1982) *Biochemical Applications of Raman and Resonance Raman Spectroscopies.* Academic Press, New York.

Carey, P.R., Fast, P., Kaplan, H. and Pozsgay, M. (1986) Molecular structure of the protein crystal from *Bacillus thuringensis*: a Raman spectroscopic study. *Biochim. Biophys. Acta* **872**: 169–76.

Carmona, P., Ramos, J.M., de Cózar, M. and Monreal, J.P. (1987) Conformational features of lipids and proteins in myelin membranes using Raman and infrared spectroscopy. *J. Raman Spectrosc.* **18**:473–76.

Caswell, D.S. and Spiro, T.G. (1986) Tyrosine and tryptophan modification monitored by ultraviolet resonance Raman spectroscopy. *Biochim. Biophys. Acta* **873**:73–78.

Chen, W., Nie, S., Kuck, J.F.R. Jr and Yu, N.-T. (1991) Near-infrared Fourier transform Raman and conventional Raman studies of calf γ-crystallins in the lyophilized state and in solution. *Biophys. J.* **60**:447–55.

Chumanov, G.D., Efremov, R.G. and Nabiev, I.R. (1990) Surface-enhanced Raman spectroscopy of biomolecules. Part I. Water-soluble proteins, dipeptides and amino acids. *J. Raman Spectrosc.* **21**:43–48.

Clark, A.H., Saunderson, D.H.P. and Suggett, A. (1981) Infrared and laser-Raman spectroscopic studies of thermally-induced globular protein gels. *Int. J. Peptide Protein Res.* **17**:353–64.

deNoyer, L.K. and Dodd, J.G. (1990) Maximum likelihood smoothing of noisy data. *Amer. Laboratory* **22 (5)**:21–27.

deNoyer, L.K. and Dodd, J.G. (1991) Maximum likelihood deconvolution for spectroscopy and chromatography. *Amer. Laboratory* **23 (12)**:24D–24H.

Frushour, B.G. and Koenig, J.L. (1975) Raman spectroscopy of proteins, in *Advances in Infrared and Raman Spectroscopy*, Vol. 1 (eds R.J.H. Clark and R.E. Hester), Heyden and Son, London, pp. 35–97.

Gerrard, D.L. and Bowley, H.J. (1988) Raman spectroscopy. *Anal. Chem.* **60**:368R–377R.

Góral, J. and Zichy, V. (1990) Fourier transform Raman studies of materials and compounds of biological importance. *Spectrochimica Acta* **46A**:253–275.

Hamodrakas, S.J., Kamitsos, E.I. and Papadopoulou, P.G. (1987) Laser Raman and infrared spectroscopic studies of protein conformation in the eggshell of the fish *Salmo gairdneri*. *Biochim. Biophys. Acta* **913**:163–69.

Harada, I., Takamatsu, T., Tasumi, M. and Lord, R.C. (1982) Raman spectroscopic study of the interaction between sulfate anion and an imidazolium ring in ribonuclease A. *Biochemistry* **21**:3674–77.

Harada, I. and Takeuchi, H. (1986) Raman and ultraviolet resonance Raman spectra of proteins and related compounds, in *Spectroscopy of Biological Systems*, (eds R.J.H. Clark and R.E. Hester), John Wiley and Sons, New York, pp. 113–75.

Hirotsuka, M. and Nakai, S. (1990) Raman spectroscopic study of soy protein-soy lecithin complex. *1990 IFT Annual Meeting Abstracts*, No. 85, p. 119.

Honzatko, R.B. and Williams, R.W. (1982) Raman spectroscopy of avidin: secondary structure, disulfide conformation, and the environment of tyrosine. *Biochemistry* **21**:6201–05.

Hudson, B. and Mayne, L. (1986) Ultraviolet resonance Raman spectroscopy of biopolymers. *Methods Enzymol.* **130**:331–50.

Itoh, K., Osaki, Y., Mizuno, A. and Iryama, K. (1983) Structural changes in the lens proteins of hereditary cataracts monitored by Raman spectroscopy. *Biochemistry* **22**:1773–78.

Kitagawa, T., Azuma, T. and Hamaguchi, K. (1979) The Raman spectra of Bence-Jones proteins. Disulfide stretching frequencies and dependence of Raman intensity of tryptophan residues on their environments. *Biopolymers* **18**:451–61.

Krimm, S. (1987) Peptides and proteins, in *Biological Applications of Raman Spectroscopy*, (ed. T.G. Spiro), John Wiley and Sons, New York, pp. 1–45.

Krimm, S. and Bandekar, J. (1986) Vibrational spectroscopy and conformation of peptides, polypeptides and proteins. *Adv. Prot. Chem.*, **38**:181–364.

Larsson, K. (1976) Raman spectroscopy for studies of interface structure in aqueous dispersions, in *Lipids, Vol. 2: Technology*, (eds R. Paoletti, G. Jacini and R. Porcellati), Raven Press, New York, pp. 355–60.

Larsson, K. and Rand, R.P. (1973) Detection of changes in the environment of hydrocarbon chains by Raman spectroscopy and its application to lipid–protein systems. *Biochim. Biophys. Acta* **326**:245–55.

Levin, I.W. and Lewis, E.N. (1990) Fourier transform Raman spectroscopy of biological materials. *Anal. Chem.* **62**:1101A–11A.

Li-Chan, E. and Nakai, S. (1991a) Importance of hydrophobicity of proteins in food emulsions, in *Microemulsions and Emulsion in Foods*, (eds M. El-Nokaly and D. Cornell), ACS Symposium Series No. 448, American Chemical Society, Washington DC, pp. 193–212.

Li-Chan, E. and Nakai, S. (1991b) Raman spectroscopic study of thermally and/or dithiothreitol induced gelation of lysozyme. *J. Agric. Food Chem.* **39**:1238–45.

Lin, V.J.C. and Koenig, J.L. (1976) Raman studies of bovine serum albumin. *Biopolymers* **15**:203–18.

Lippert, J.L., Lindsay, R.M. and Schultz, R. (1981) Laser Raman characterization of conformational changes in sarcoplasmic reticulum induced by temperature, Ca^{2+} and Mg^{2+}. *J. Biol. Chem.* **256**:12411–16.

Lippert, J.L., Tyminski, D. and Desmeules, P.J. (1976) Determination of the secondary structure of proteins by laser Raman spectroscopy. *J. Amer. Chem. Soc.* **98**:7075–80.

Lord, R.C. and Yu, N.-T. (1970) Laser-excited Raman spectroscopy of biomolecules. I. Native lysozyme and its constituent amino acids. *J. Mol. Biol.* **50**:509–24.

Maxfield, F.R. and Scheraga, H.A. (1977) A Raman spectroscopic investigation of the disulfide conformation in oxytocin and lysine vasopressin. *Biochemistry* **16**:4443–49.

Melander, W. and Horvath, C. (1977) Salt effects on hydrophobic interactions in precipitation and chromatography of proteins: an interpretation of the lyotropic series. *Arch. Biochem. Biophys.* **183**:200–15.

Mikkelsen, R.B., Verma, S.P. and Wallach, D.F.H. (1978) Effect of transmembrane ion gradients on Raman spectra of sealed hemoglobin-free erythrocyte membrane vesicles. *Proc. Natl. Acad. Sci. USA* **75**:5478–82.

Miura, T., Takeuchi, H. and Harada, I. (1991) Raman spectroscopic characterization of tryptophan side-chains in lysozyme bound to inhibitors: role of the hydrophobic box in the enzymatic function. *Biochemistry* **30**: 6074–80.

Nabiev, I.R., Chumanov, G.D. and Efremov, R.G. (1990) Surface-enhanced Raman spectroscopy of biomolecules. Part II. Application of short- and long-range components of SERS to the study of the structure and function of membrane proteins. *J. Raman Spectrosc.* **21**:49–53.

Nakai, S., Li-Chan, E., Hirotsuka, M., Vazquez, M.C. and Arteaga, G. (1991) Quantitation of hydrophobicity for elucidating the structure-activity relationships of food proteins, in *Interactions of Food Proteins*, (eds N. Parris and R. Barford), ACS Symposium Series No. 454, American Chemical Society, Washington DC, pp. 42–58.

Nakanishi, M., Takesada, H. and Tsuboi, M. (1974) Conformation of the cystine linkages in bovine α-lactalbumin as revealed by its Raman effect. *J. Mol. Biol.* **89**:241–43.

Nemoto, N., Kiyohara, T., Tsunashima, Y. and Kurata, M. (1983) Raman, IR and CD spectroscopy on aggregates and gels of lysozyme in alcohols. *Bull. Inst. Chem. Res.* **61**:203–13.

Ni, F. and Scheraga, H.A. (1985) Resolution enhancement in spectroscopy by maximum entropy Fourier self-deconvolution, with applications to Raman spectra of peptides and proteins. *J. Raman Spectrosc.* **16**:337–49.

Ozaki, Y. (1988) Medical application of Raman spectroscopy. *Appl. Spectrosc. Rev.* **24**:259–312.

Painter, P.C. (1984) The application of Raman spectroscopy to the characterization of food, in *Food Analysis. Principles and Techniques. Volume 2. Physicochemical Techniques*, (eds D.W. Gruenwedel and J.R. Whitaker), Marcel Dekker, New York, pp. 511–45.

Painter, P.C. and Koenig, J.L. (1976). Raman spectroscopic study of the proteins of egg white. *Biopolymers* **15**:2155–66.

Pande, J., McDermott, M.J., Callender, R.H. and Spectro, A. (1989) Raman spectroscopic evidence for a disulfide bridge in calf γ_{II} crystallin. *Arch. Biochem. Biophys.* **269**:250–55.

Prescott, B., Renugopalakrishnan, B., Glimcher, M.J., Bhushan, A. and Thomas, G.J. Jr. (1986) A Raman spectroscopic study of hen egg yolk phosvitin: structures in solution and in the solid state. *Biochemistry* **25**:2792–98.

Przybycien, T.M. and Bailey, J.E. (1989) Structure–function relations in the inorganic salt-induced precipitation of α-chymotrypsin. *Biochim. Biophys. Acta* **995**:231–45.

Przybycien, T.M. and Bailey, J.E. (1991) Secondary structure perturbations in salt-induced protein precipitates. *Biochim. Biophys. Acta* **1076**:103–11.

Raman, C.V. (1928) A new radiation. *Indian J. Phys.* **2**:387–98.

Raman, C.V. and Khrishnan, K.S. (1928) A new type of secondary radiation. *Nature* **121**:501–2.

Schrader, B., Hoffman, A., Simon, A. and Sawatzki, J. (1991) Can a Raman renaissance be expected via the near-infrared Fourier transform technique? *Vibrational Spectroscopy* **1**:239–50.

Siamwiza, M.N., Lord, R.C., Chen, M.C., Takamatsu, T., Harada, I., Matsuura, H. and Shimanouchi, T. (1975) Interpretation of the doublet at 850 and 830 cm⁻¹ in the Raman spectra of tyrosyl residues in proteins and certain model compounds. *Biochemistry* **14**:4870–76.

Strommen, D.P. and Nakamoto, K. (1984) *Laboratory Raman Spectroscopy*. John Wiley and Sons, New York.

Sugeta, H., Go., A. and Miyazawa, T. (1972) S–S and C–S stretching vibrations and molecular conformations of dialkyl disulfides and cystine. *Chem. Lett.* **1972**:83–86.

Sugeta, H., Go, A. and Miyazawa, T. (1973) Vibrational spectra and molecular conformations of dialkyl disulfides. *Bull. Chem. Soc. Jpn.* **46**:3407–11.

Susi, H. and Byler, D.M. (1983). Protein structure by Fourier transform infrared spectroscopy: second derivative spectra. *Biochem. Biophys. Res. Comm.* **115**:391–97.

Susi, H. and Byler, D.M. (1987) Fourier transform infrared study of proteins with parallel β-chains. *Arch. Biochem. Biophys.* **258**:465–69.

Susi, H. and Byler, D.M. (1988a) Fourier transform infrared spectroscopy in protein conformation studies, in *Methods for Protein Analysis*, (eds J.P. Cherry and R.A. Barford), American Oil Chemists' Society, Champaign, III, pp. 235–55.

Susi, H. and Byler, D.M. (1988b) Fourier deconvolution of the amide I Raman band of proteins as related to conformation. *Appl. Spectrosc.* **42**:819–26.

Tu, A.T. (1986) Peptide backbone conformation and microenvironment of protein side-chains, in *Spectroscopy of Biological Systems*, (eds R.J.H. Clark and R.E. Hester), John Wiley and Sons, New York, pp. 47–112.

Van Dael, H., Lafaut, J.P. and Van Cauwelaert, F. (1987) Tyrosine group behaviour in bovine α-lactalbumin as revealed by its Raman effect. *Eur. Biophys. J.* **14**:409–14.

Van de Ven, M., Meijer, J., Verwer, W., Levine, Y.K. and Sheridan, J.P. (1984) Derivative Raman spectroscopy applied to biomembrane systems. *J. Raman Spectrosc.* **15**:86–89.

Van Wart, H.E., Lewis, A., Scheraga, H.A. and Saeva, F.D. (1973) Disulfide bond dihedral angles from Raman spectroscopy. *Proc. Natl. Acad. Sci. USA* **70**:2619–23.

Verma, S.P. and Wallach, D.F.H. (1977a) Changes of Raman scattering in the CH-stretching region during thermally induced unfolding of ribonuclease. *Biochem. Biophys. Res. Comm.* **74**:473–79.

Verma, S.P. and Wallach, D.F.H. (1977b) Raman spectra of some saturated, unsaturated and deuterated C₁₈ fatty acids in the HCH-deformation and CH-stretching regions. *Biochim. Biophys. Acta* **486**:217–27.

Vogel, H. and Jähnig, F. (1986) Models for the structure of outer-membrane proteins of *Escherichia coli* derived from Raman spectroscopy and prediction methods. *J. Mol. Biol.* **190**:191–99.

Williams, R.W. (1981) Determination of the secondary structure of proteins from the amide I band of the laser Raman spectrum. *J. Mol. Biol.* **152**:783–813.

Williams, R.W, (1983) Estimation of protein secondary structure from the laser Raman amide I spectrum. *J. Mol. Biol.* **166**:581–603.

Williams, R.W. (1986) Protein secondary structure analysis using Raman Amide I and Amide III spectra. *Methods Enzymol.* **130**:311–31.

Yada, R.Y., Jackman, R.L. and Nakai, S. (1988) Secondary structure prediction and determination of proteins—a review. *Int. J. Peptide Protein Res.* **31**:98–108.

Yager, P. and Gaber, B.P. (1987) Membranes, in *Biological Applications of Raman Spectroscopy*, Vol. 1, (ed. T.G. Spiro), John Wiley and Sons, New York, pp. 204–61.

Yu, N.-T. (1977) Raman spectroscopy: A conformational probe in biochemistry. *CRC Crit. Rev. Biochem.* **4**:229–280.

Yu, N.T., Jo, B.H. and Liu, S.A. (1972). A laser Raman spectroscopic study of the effect of solvation on the conformation of ribonuclease A. *J. Amer. Chem. Soc.* **94**:7572–75.

Index